BASIC ASTRONOMY

BASIC ASTRONOMY

WITH PROJECTS FOR AMATEURS AND STUDENTS

By

H. HAYSHAM

EXTRA MASTER MARINER

Fellow of the Royal Astronomical Society

Lecturer in Navigation and Astronomy

at the

South Shields Marine and Technical College

PUBLISHED BY THOMAS REED PUBLICATIONS LIMITED

SUNDERLAND AND LONDON

First Edition - 1971

PRINTED BY THOMAS REED AND COMPANY LIMITED
SUNDERLAND, GLASGOW AND LONDON

Comet Bennett 1969i

This spectacular comet, discovered by J. C. Bennett, Pretoria, became a naked-eye object to observers in both northern and southern hemispheres. The above photograph was taken at 04 h 05 m universal time on 5 April, 1970 by FRANK J. ACFIELD, FRAS *from his observatory at Forest Hall, Newcastle upon Tyne.*

FOREWORD

This book is directed towards the growing number of people now taking a deeper interest in astronomy, either as amateurs or students.

In my own experience, both as an amateur astronomer and as a lecturer, it seems to me that the needs of the new recruit to the amateur ranks and the ordinary level student are about the same; each requires a basic theoretical and practical knowledge of the subject. In order to meet these requirements each chapter is divided into two sections: the first part provides the theory, while the second part is devoted to related projects and research problems. Nearly 60 varied projects are included in the text so that there is ample scope for those who possess binoculars or a small telescope, as well as those who have no such instruments at their disposal.

The text adequately covers the G.C.E. Ordinary Level syllabuses of the London University and the Associated Examining Board. Incidentally, I know of many amateur astronomers, young and old, who have derived great pleasure and satisfaction from passing the examination at "O" level, and I hope that this book may induce others to avail themselves of the opportunity of using their hobby as a means of attaining a worthy academic standard.

I have endeavoured to include project material suitable for southern observers in addition to those who live north of the equator, and I hope this will appeal not only to our cousins in the antipodes but also to young navigators whose voyages afford them a unique opportunity of seeing the splendours of the heavens from all parts of the globe.

I wish to express my gratitude to the members of the South Shields Astronomical Society upon whose wealth of experience I have made frequent calls during the preparation of material for this book. Edward Duncan, Aileen Thompson, Edward Mitchinson and the late Stanley Buglass were particularly helpful in reading and commenting upon much of the subject matter.

I am especially grateful to William Embleton, O.B.E., who has acted as Technical Editor. His advice and assistance in the presentation of the manuscript and in the reading and correcting of the proofs have been invaluable.

I am also indebted to those persons and institutions who have generously supplied the photographs which I am pleased to acknowledge where they appear in the text.

Cleadon, Co. Durham, Harry Haysham.

CONTENTS

LIST OF PROJECTS AND RESEARCH

LIST OF TABLES

LIST OF PLATES

CHAPTER 1

ASPECTS OF THE SKY

An observer viewing the beauty of the heavens on a clear night may well imagine the stars to be set on the inner surface of a vast hollow dome. Fortunately, this perfectly natural impression is quite useful because it is often convenient to think of heavenly bodies as being situated on the inside of a sphere which is known to astronomers as the *celestial sphere*. It is a purely imaginary sphere of no definite radius, since in reality there are considerable differences in the distances of the stars from the earth. However, before turning our thoughts skywards we must first do a little ground work.

CIRCLES OF THE SPHERE

Circles on a sphere are of two main types: *great circles* and *small circles*. A great circle is one whose plane passes through the centre of the sphere, whilst a small circle is one whose plane does not. For example, if a nice round orange is cut into two equal parts, the knife must have passed through its centre. The circumference of the cut represents a great circle and the flat surface the *plane* of this great circle. Since a cut made in any direction would divide the orange into two equal portions, it follows that any number of great circles may be drawn on a sphere. Had the knife not passed through the centre of the orange the result would have been two unequal portions, with the surface cut representing a small circle. From any point on a great circle, the line joining it to the centre of the sphere is a radius. The angle contained between the radii of two different points is a measure of the distance along the arc of the great circle passing through the points or, in other words, the *angular distance* between them. Angular measurements are frequently met with in astronomy and the units involved are degrees, minutes and seconds.

$$1 \text{ degree } (1^\circ) = \frac{1}{360} \text{ circle}$$

$$1 \text{ minute } (1') = \frac{1}{60} \text{ degree}$$

$$1 \text{ second } (1'') = \frac{1}{60} \text{ minute}$$

Decimal figures are also used: thus 10° 20′ 30″ may be stated as 10° 20′·5, and 10° 24′ as 10°·4.

THE TERRESTRIAL SPHERE

The earth is a spinning body and the points where the axis of rotation meet the surface are known as the *north* and *south poles.* Midway between the poles lies a great circle called the *equator;* it divides the earth into two equal hemispheres, north and south, and all points on it are 90° from each of the poles. Small circles lying parallel to the equator, and diminishing in size with increasing distance from it, are known as *parallels of latitude.* Lines of *longitude*, or *meridians*, are half great circles joining the poles, which means that at any place the meridian represents the north-south direction. The one passing through the meridian instrument at the original site of the Royal Greenwich Observatory in London is taken as the *prime meridian* for the purpose of measuring longitude. Because of the poor observing conditions prevailing in London, the Royal Observatory was moved to a new location at Herstmonceux Castle in Sussex.

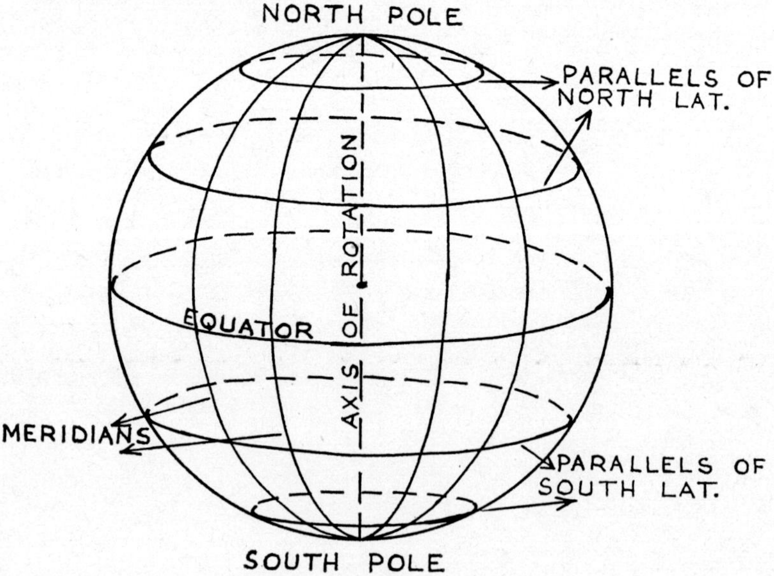

Fig. 1.1
Circles of the Terrestrial Sphere.

The *latitude* of a point on the earth's surface is the arc of a meridian measured north or south from the equator to the parallel of latitude through the point. Its value ranges from 0° at the equator to 90° at the poles, north and south latitudes being indicated by the abbreviations N and S respectively. The *longitude* of a point is the arc of the equator, or the angle at the pole, between the Greenwich meridian and the meridian passing through the point, its value ranging from 0° at the Greenwich meridian to 180° east and west from Greenwich. The abbreviations E and W indicate east and west longitudes respectively. In order to define the position of any place on the earth it is necessary to know both its latitude and longitude. Assuming the earth to be a perfect sphere,

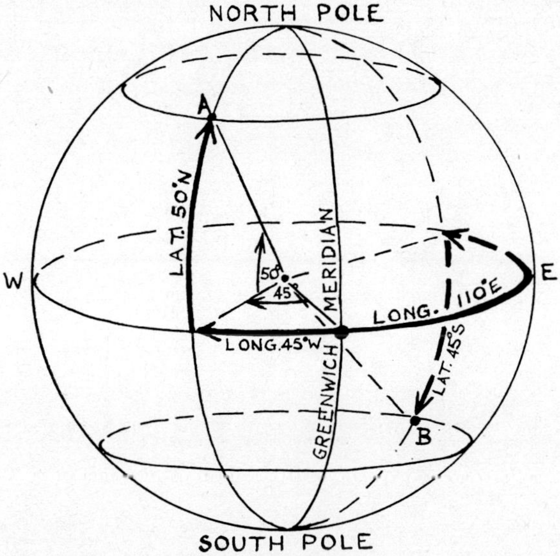

Fig. 1.2
Latitude and Longitude.

the arcs defining latitude and longitude are equal to the angles subtended by them at the earth's centre.

THE CELESTIAL SPHERE

Only half of the celestial sphere can be seen at any given time, but due to the earth's rotation the visible half is constantly changing. All measurements upon it are made along arcs of great circles and are, therefore, of an angular nature. It is useful to remember that the angular diameters of the sun and moon are each about $\frac{1}{2}°$; the two pointer stars of the well known constellation of The Plough are just over 5° apart. The position of an object on the celestial sphere is defined by two angular measurements, just as that of a point on the earth's surface is given by the two angular measurements of latitude and longitude.

AZIMUTH AND ALTITUDE

To any observer the *zenith* is the point on the celestial sphere directly overhead; the opposite point is called the *nadir*. The great circle halfway between the zenith and the nadir is the *celestial horizon*, the horizon of astronomy. It is more nearly coincident with the sea horizon than with the often irregular skyline of the land. All points on the celestial horizon are 90° from the zenith. Half great circles joining the zenith and the nadir are called *vertical circles*. They may be readily imagined by pointing upwards

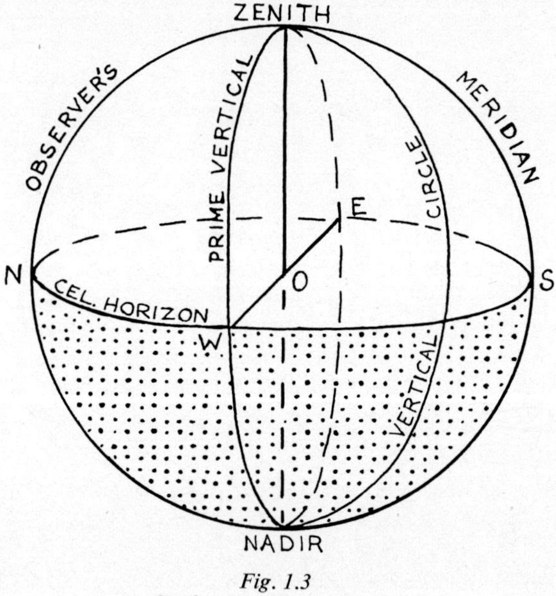

Fig. 1.3
Circles through the Zenith.

to the zenith and then, whilst facing in a different direction each time with the arm fully extended, describing arcs vertically downwards against the background of the sky to cross the horizon at right angles. The one joining the east and west points of the horizon (E and W in fig. 1.3) is called the *prime vertical circle.*

The one joining the north and south points (N and S in fig. 1.3) is most important because it coincides with the observer's terrestrial meridian expanded outwards on to the celestial sphere. It is known as the *observer's celestial meridian* or, more simply, "*the meridian*".

At any instant of time an observer may refer to the position of a star, or some other celestial body, by two co-ordinates called *azimuth* and *altitude*. Azimuth is the arc of the celestial horizon from the north point eastwards to the vertical circle through the body. Thus, due north is 000° (or 360°); due east 090°; due south 180°; and due west 270°. Altitude is the vertical angle between the celestial horizon and the star, *i.e.* an arc of its vertical circle. Its value ranges from 0° at the horizon to 90° at the zenith. The arc from the zenith to the star is aptly named the *zenith distance* and, because from zenith to horizon is 90°, it follows that:

$$\textit{Zenith Distance} = 90° - \textit{Altitude}.$$

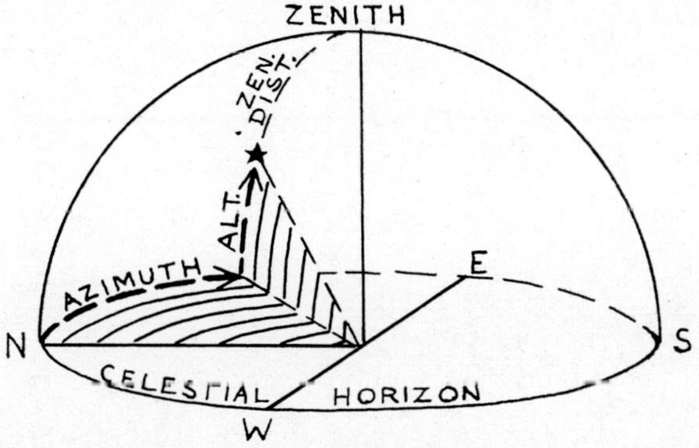

Fig. 1.4
Azimuth and Altitude.

The altitude-azimuth system of indicating position is a convenient one because it is fairly simple to visualise the angular measurements involved. Its main disadvantage arises from the fact that the rotation of the earth on its axis imparts an apparent rotation to the celestial sphere which, in turn, causes a steady change in the values of altitude and azimuth of any given object. This can be verified by noting, at half-hourly intervals, the position of the sun or a star with reference to a fixed object such as a chimney.

CELESTIAL POLES AND CELESTIAL EQUATOR

The earth's axis of rotation produced in both directions meets the celestial sphere at two points, called the *north* and *south celestial poles.* The earth's equator expanded outwards on to the celestial sphere gives the *celestial equator*, or *equinoctial.* Half great circles joining the celestial poles are called *hour circles.* It was mentioned earlier that the observer's terrestrial meridian expanded outwards on to the celestial sphere gives the observer's celestial meridian, a most important reference circle in the sky: the half joining the north and south celestial poles through the zenith is known as the *superior meridian*; the other half which passes through the nadir is the *inferior meridian.* Fig. 1.5 represents the celestial sphere for an observer in a middle-northern

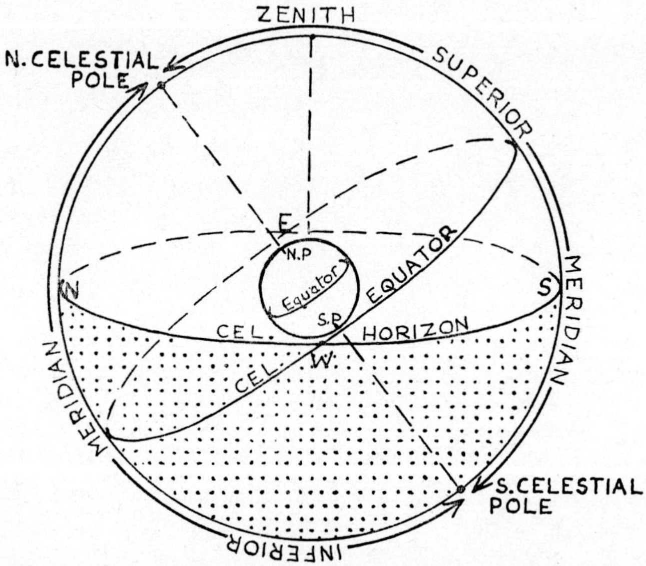

Fig. 1.5
Points and Circles on the Celestial Sphere related to those on Earth.

latitude. It has been drawn with zenith uppermost so that the reader may use the diagram to help visualise the various points and circles as they appear in the sky. A simplified form of the diagram is shown in fig. 1.6 in which the earth has been omitted for clarity, and the observer is shown at the centre of the celestial sphere. This may seem a drastic step until it is remembered the celestial sphere is so vast that by comparison the earth's size is insignificant—

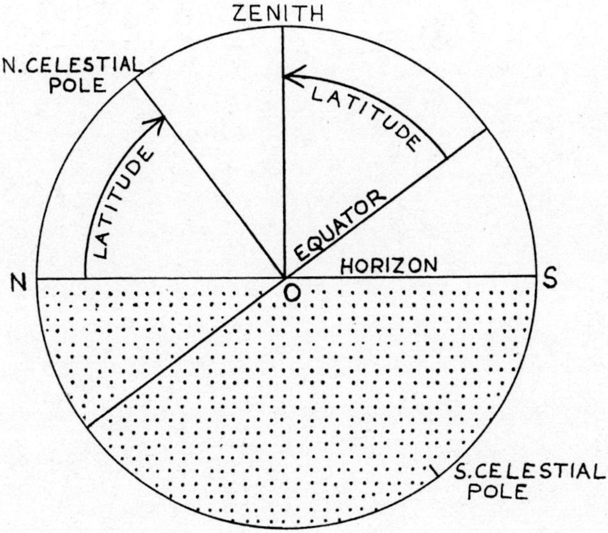

Fig. 1 6
The Celestial Sphere in the Plane of the Observer's Meridian.

much like the smallest needle in the largest haystack. Examination of these two figures will yield the following useful information:

(a) The celestial equator crosses the horizon at the east and west points.

(b) The arc of the observer's celestial meridian from the celestial equator to his zenith is equal to the arc of his terrestrial meridian from the equator to his position, *i.e.* his latitude.

(c) In the observer's meridian the celestial equator is inclined to the horizon at an angle equal to the observer's *co-latitude, i.e.* (90°—latitude).

(d) The north celestial pole is above the horizon of an observer in north latitude, the south celestial pole being below the horizon. The reverse is true for an observer in the southern hemisphere.

(e) The altitude of the elevated pole equals the observer's latitude.

Observers in the northern hemisphere are very fortunate in having the north celestial pole marked by the bright star *Polaris* or, to give it its constellation name, *Alpha Ursa Minoris.* It is easily located by means of the *pointer stars,* Dubhe and Merak, of the conspicuous northern constellation of The Plough, as shown in fig. 1.7. At present, Polaris is only about 1° from the north

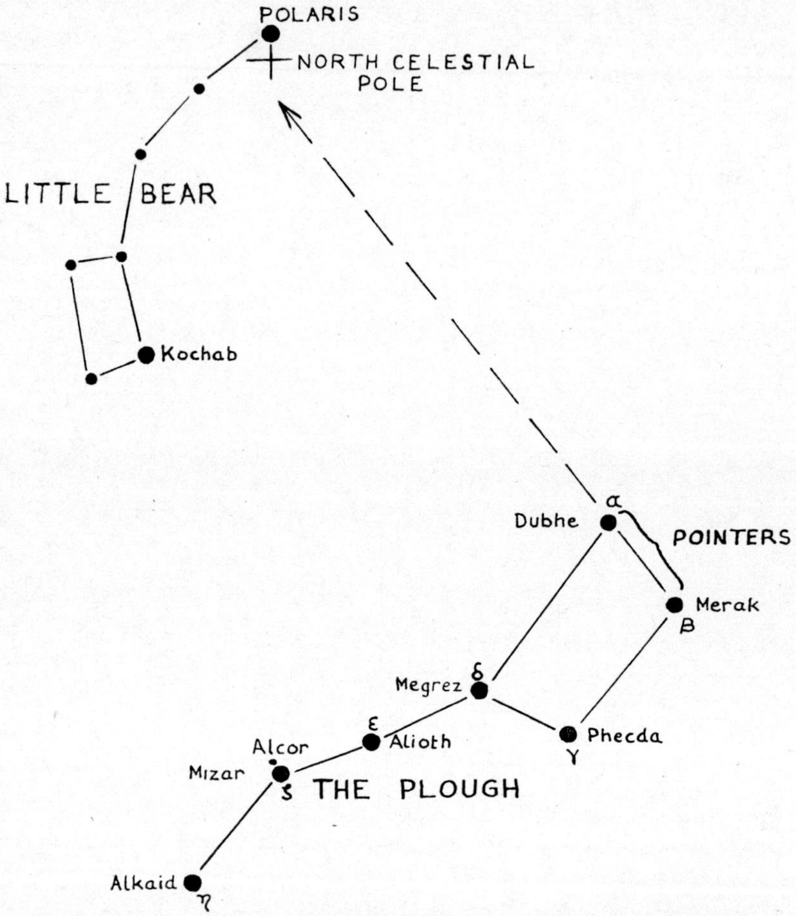

Fig. 1.7
The Plough or Big Dipper.

celestial pole, so that its altitude gives a good approximation of the observer's latitude. Also, its direction is very nearly true north. Southern observers are less fortunate, for there is no bright star sufficiently close to the south celestial pole to serve as a pole star.

RIGHT ASCENSION AND DECLINATION

The disadvantages of the altitude-azimuth system of indicating position on the celestial sphere caused astronomers to derive a system having no reference to the observer. The co-ordinates employed, called *Right Ascension* (*R.A.*) and *declination* (*decl.*), are very much akin to latitude and longitude used on earth. The Right Ascension of star is the angular distance from the *First Point of Aries* (symbol ♈) along the celestial equator to the hour circle through the star. Its value is usually expressed in time from 0 hours at Aries eastwards to 24 hours at Aries again. As we shall see shortly, the First Point of Aries, or Vernal Equinox, is the point where the sun crosses the celestial equator on or about March 21. The declination of a star is its angular distance north or south of the celestial equator. Its value ranges from 0° at the celestial equator to 90° at the celestial poles, north and south declinations being indicated by the abbreviations N and S, or plus

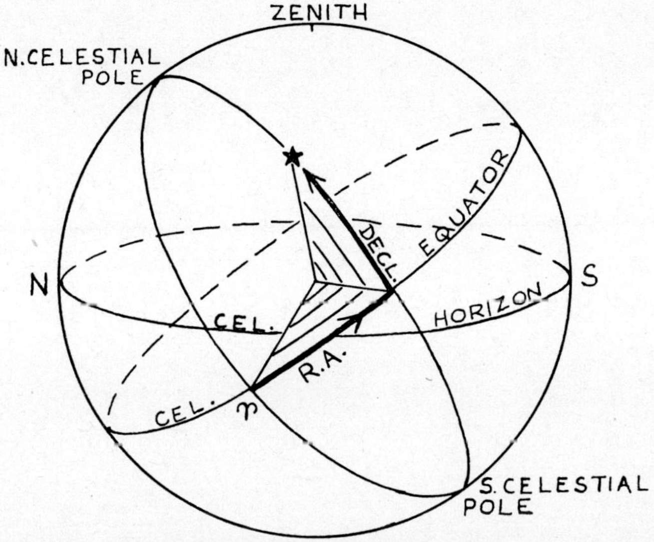

Fig. 1.8
Right Ascension and Declination.

and minus signs respectively. In catalogues and on maps the positions of stars and other celestial bodies are indicated by these co-ordinates.

DIURNAL MOTION OF THE STARS

The daily rotation of the earth on its axis from west to east manifests itself as a motion of the celestial sphere in the opposite direction about an axis joining the north and south celestial poles, imparting to all celestial bodies an *apparent diurnal motion* across the sky from east to west. The paths traced out by the stars on the celestial sphere during the course of the day, called *diurnal circles*, are parallel to the celestial equator. Although the diurnal motion of the sky changes with change of latitude, the celestial poles have no movement since they are the points about which the celestial sphere is apparently rotating.

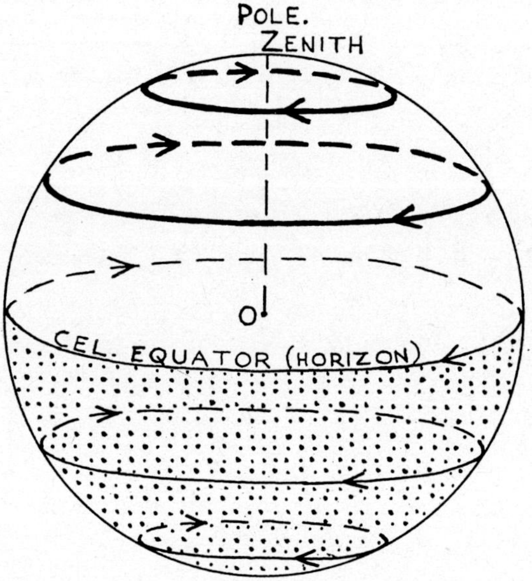

Fig. 1.9
Diurnal Circles Observed at the North Pole.

To an observer at either of the earth's poles the elevated pole is in the zenith; the celestial equator coincides with the horizon, and the diurnal circles, being parallel to the celestial equator, are also parallel to the horizon. Therefore, at the north pole, an observer will see all the stars having north declination but none of those having south declination. Conversely, an observer at the south pole will see all the southern stars but no northern stars. In either case the visible stars will move around the sky at fixed altitudes, never setting.

At the equator, the celestial poles are on the horizon and the celestial equator passes through the zenith. The stars, in tracing out their diurnal paths, rise and set vertically and remain above

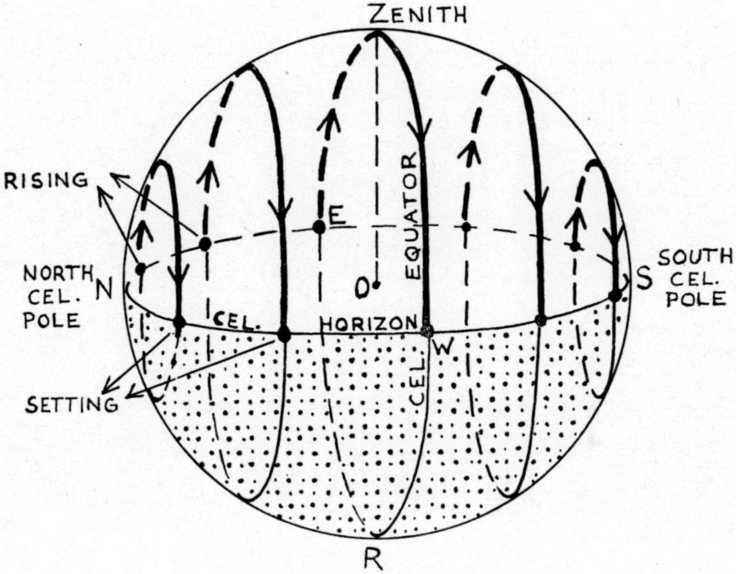

Fig. 1.10
Diurnal Circles Observed at the Equator.

the horizon for half a day. It is possible for an equatorial observer to see all parts of the celestial sphere, although only half of it is visible at any instant of time.

To an observer in either hemisphere at a position intermediate between the pole and the equator, the diurnal circles are oblique. The celestial pole is elevated above the horizon to an angle equal to the latitude, and the celestial equator to an angle in the meridian equal to the co-latitude. For example, in lat. 50° N, the celestial pole lies 50° above the north horizon and the celestial equator 40° above the south horizon. Stars of south declination greater than 40° have their diurnal circles completely below the horizon and will never rise. Near the south horizon the visible arcs of the diurnal circles are relatively small, but for the stars whose rising points are further north the visible arcs become progressively larger and the stars remain above the horizon longer. Stars of north declination greater than 40° will have their diurnal circles completely above the horizon and never set; such stars are said to be *circumpolar*.

In following its diurnal path across the sky, a star attains its maximum altitude when it is on the observer's superior meridian, at which time it is said to be at *upper meridian passage*, or *upper*

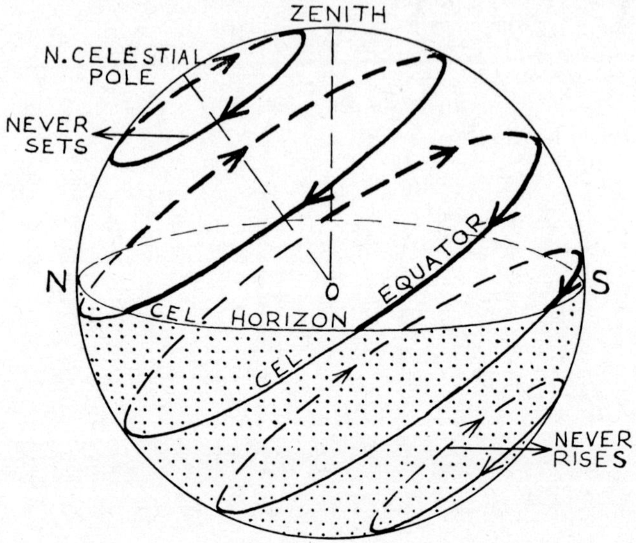

Fig. 1.11
Diurnal Circles Observed at an Intermediate Latitude.

transit. At this time it is conveniently placed for observation in order to determine latitude by the method outlined in the practical section at the end of this chapter. In the case of a circumpolar star, the whole of the diurnal circle is above the horizon and the star is visible when it is on the observer's inferior meridian. This occurs about 12 hours after upper transit and the star is then said to be at *lower meridian passage*, or *lower transit*.

ANNUAL PATH OF THE SUN—THE ECLIPTIC

In addition to rotating on its axis, the earth performs another motion—that of revolving around the sun. It takes the earth a year to complete one revolution, and in so doing it causes the sun to appear to move around the celestial sphere. The stars are not visible in the daytime because they are drowned in the strong light of the sun, but if they could be seen, how would the sun move with respect to the stars? To answer this question we must bear in mind that the stars are at such immense distances from us they may be considered as fixed points deep in space. The sun, however, at 93 000 000 miles (150 000 000 km) from us is only a stone's throw away by astronomical standards and, to an observer on the moving earth, tends to move against the background of the stars in the same way as when travelling through the countryside the near objects tend to move against the background of those more distant. Fig. 1.12 shows the earth in various positions as it progresses in its

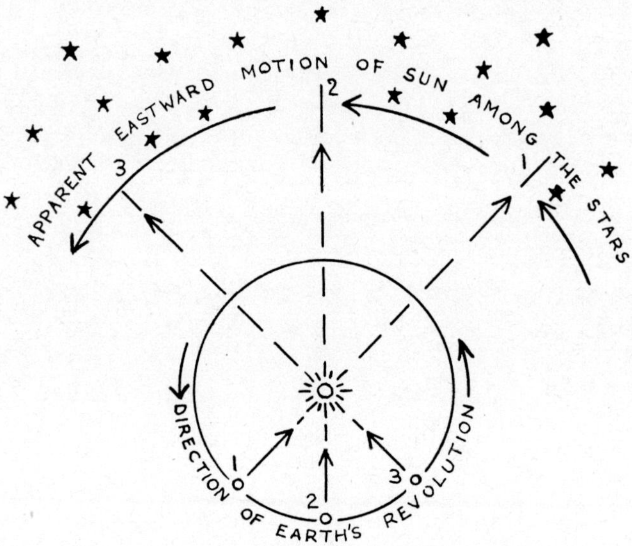

Fig. 1.12
The Apparent Eastward Motion of the Sun.

annual journey around the sun. From each position of the earth the sun appears in a different direction, so that if the stars were visible the sun would appear to have an eastwards motion amongst them, resulting in an increase in its Right Ascension. The sun would also appear to move north and south among the stars, resulting in a change in its declination, a fact which is clearly evident from the high noon-day sun in summer and the low noon-day sun in winter. If the positions of the sun on the celestial sphere were plotted successively on different dates throughout the year it would be found to follow a great circle path, inclined at an angle of nearly $23\frac{1}{2}°$ to the celestial equator and known as the *ecliptic.*

The inclination of the ecliptic is due to the earth's equator being inclined at nearly $23\frac{1}{2}°$ to the plane of its orbit.

The point where the sun crosses the celestial equator from south to north, on or about March 21, is the *Spring or Vernal Equinox.* It is also known as the First Point of Aries which, as we saw earlier, is the reference point for Right Ascension. After the Spring

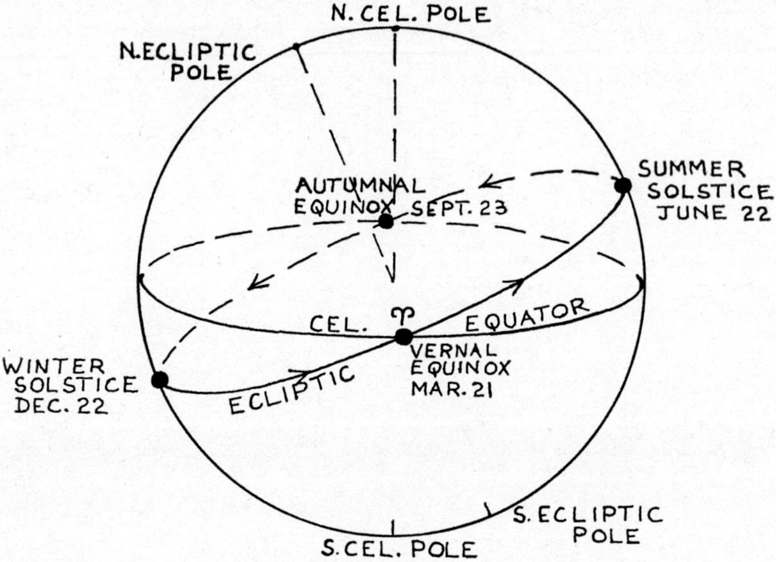

Fig. 1.13
The Ecliptic.

Equinox, the sun enters the north celestial hemisphere where it remains for the next six months, reaching a maximum north declination of nearly 23½° on or about June 22 at the *Summer Solstice*, and returning to the celestial equator at the *Autumnal Equinox* about September 23. It then enters the south celestial hemisphere for six months, attaining a maximum south declination of nearly 23½° about December 22 at the *Winter Solstice*, returning to the Spring Equinox about March 21. The relation between the solstices, equinoxes and the seasons is discussed more fully in Chapter 3.

THE ZODIAC

The *Zodiac* is a band around the celestial sphere bounded by two circles 8° on either side of the ecliptic. It is divided into 12 zones, each 30° in length, called the *Signs of the Zodiac*. They gained their names from the star constellations occupying the respective zones about 2300 years ago. The signs are:

Aries (Ram)		Libra (Scales)	
Taurus (Bull)	Spring	Scorpius (Scorpion)	Autumn
Gemini (Twins)		Sagittarius (Archer)	

Cancer (Crab)		Capricornus (Goat)	
Leo (Lion)	Summer	Aquarius (Water Carrier)	Winter
Virgo (Virgin)		Pisces (Fishes)	

The gravitational pull of the moon and the sun on the earth's equatorial bulge causes the earth's axis of rotation to describe a slow conical movement, called *precession.* It results in the celestial poles moving around the poles of the ecliptic, taking about 26 000 years to complete one revolution. There is also a corresponding movement of the equinoxes such that they slide gradually westwards along the ecliptic at the rate of just under 1 minute of arc each year. This small quantity has added up over the years since the Signs of the Zodiac were first given their names, and in the interval the equinoxes have moved about 32° to the west, taking the Signs away from the constellations with which they were originally associated; the First Point of Aries is now among the stars of the constellation Pisces. Although on March 21, when the sun arrives at the Spring Equinox it enters the Sign of Aries, it will not enter the constellation of Aries until almost a month later. Thus the Signs of the Zodiac and the constellations bearing the same names are entirely unconnected, and apart from their association with the sun's annual journey around the ecliptic, the Signs are of little interest to astronomers.

A star atlas always states the epoch for which it is drawn up so that corrections can be applied the R.A. and declination of stars to take account of precession. Precessional movement of the celestial poles is responsible for a change in the north polar star; by the year 2100 the pole will be only $\frac{1}{2}$° from Polaris, but by 7500 it will be well removed from it and the new Pole Star will be Alpha Cephei.

REVIEW QUESTIONS

1. Using an Ordnance Survey map of your locality, find the exact latitude and longitude of your position. Define these terms.

2. New York is in lat. 40° 40′ N, long. 73° 50′ W. Toronto lies 3° 02′ of latitude to the north of New York and 5° 35′ of longitude to the west. What is the latitude and longitude of Toronto?

3. On a suitable diagram indicate the positions of Bombay, lat. 19° N, long. 73° E; and Pitcairn Is., lat. 25° S, long. 130° W.

4. Define azimuth and altitude. State the azimuth and altitude of a star in each of the following positions:
(a) halfway between north and east and one-third the way from horizon to zenith, (b) 20° to the right of an observer facing due south and 30° from the zenith, (c) 30° to the north of west and halfway between horizon and zenith.

5. Name the two points on the celestial sphere having no apparent diurnal motion. At what points does the celestial equator cut the horizon at places other than the poles?

6. Which celestial pole is above the horizon in Europe and Australia?

7. How can the Pole Star be located? If the Pole Star is assumed to be at the north celestial pole, what would be its altitude from London in lat. 51° 30′ N. and Washington D.C. in lat. 38° 55′ N? In what latitude would you be if the Pole Star was on the horizon?

8. What would be the altitude of a star of decl. 0° when on the superior meridian of an observer in lat. 40° N?

9. Name the celestial circle or co-ordinate corresponding to each of the following definitions:
(a) angular distance measured eastwards from the First Point of Aries along the celestial equator.
(b) angular distance measured north or south of the celestial equator.
(c) the sun's apparent annual path.

10. Sketch the celestial sphere, showing on it the celestial poles, the celestial equator, the ecliptic, the positions of the stars Regulus (R.A. 10 h, decl. 12° N) and Rigel (R.A. 5 h, decl. 8° S) and the position of the sun in early February.

PROJECTS AND RESEARCH

1. FINDING THE MERIDIAN

The point on the horizon vertically below the Pole Star indicates the direction of true north, and hence the meridian. This is a useful method of establishing a sense of direction at night, but if it is desired to mark out the meridian on the ground it is more convenient to work in the day time, using the sun. In common with all other astronomical bodies, the sun shares an apparent motion across the sky from east to west caused, as we have seen, by the earth's rotation. Casual observation shows that shadows are long in the early hours after sunrise, becoming shorter as the sun climbs higher in the sky and lengthening again towards evening. The sun is highest in the sky when it is on the observer's meridian but it is difficult to tell exactly when the shortest shadows occur, so the following method is recommended.

Obtain a straight stick, sharpened to a point at the top, and erect it firmly into an area of level ground, checking the stick is vertical by means of a plumb line or spirit level. At about an hour or so before midday, observe the shadow cast by the stick and mark its extremity with a peg, a long nail, or other means. Tie a piece of string around the base of the shadow stick and, stretching the string tightly, mark with a ball pen where it meets the peg. For a while the shadow will continue to shorten and then it will be seen to lengthen again. Using the marked string, check the growing shadow until it is exactly the same length as it was earlier. Again, mark the point with a peg. Measure exactly the distance between

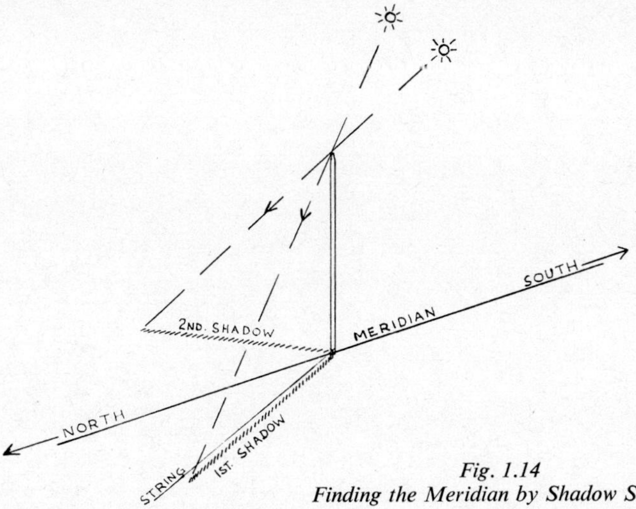

Fig. 1.14
Finding the Meridian by Shadow Stick.

the two pegs and find the central point from which a line to the base of the shadow stick indicates the meridian. The first shadow was formed at the same interval of time before meridian passage of the sun as the second shadow was formed after it. If watch times of the observations were taken then the middle of these times would be the watch time of the sun's transit across the observer's meridian. More than likely this time would not be 12 o'clock, but there is no need to worry because civil time is not provided by the sun we see. (The whole question of time will be discussed in Chapter 4.)

More than one pair of shadows could be measured on the same day, but even so the experiment should be repeated on several occasions. Once the meridian has been established it should be extended and marked out permanently as it will be found useful in further projects. Other directions such as east and west, north east, south east, south west and north west could also be marked out. Alternatively, a large flat board carrying a circular scale graduated 0° to 360° could be mounted with the 0°-180° line in the meridian. A pointer pivoted at the centre of the scale and fitted with two sight vanes would provide a useful means of measuring azimuth.

2. MEASURING ALTITUDE

A simple device for measuring altitude may be constructed as follows, with reference to fig. 1.15. Obtain a wood batten 2 in by

Fig 1.15.
Device for Measuring Altitude.

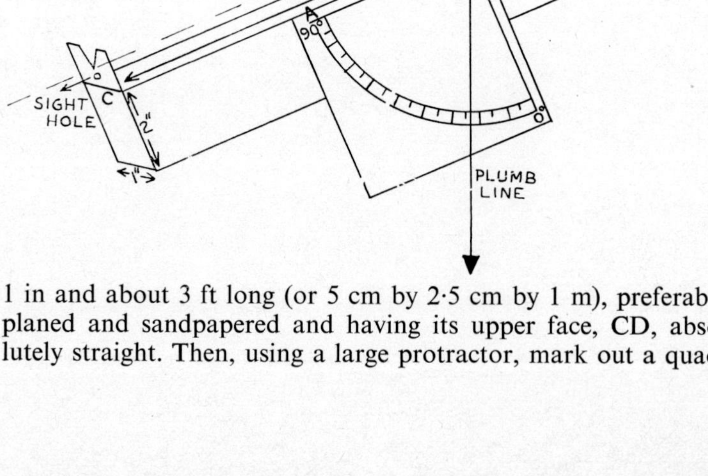

1 in and about 3 ft long (or 5 cm by 2·5 cm by 1 m), preferably planed and sandpapered and having its upper face, CD, absolutely straight. Then, using a large protractor, mark out a quad-

rant (90°) on stiff card or drawing paper and graduate it accurately in degrees, inserting further subdivisions if possible. Now glue the graduated scale on to a piece of hardboard or plywood and attach it to the batten so that the straight edge of the scale, AB, is perfectly parallel to the straight face of the batten, CD. The graduated scale could be marked directly on to the hardboard; an even simpler construction would be to fit a large protractor directly on to the batten. Insert a panel pin at the point B and suspend from it a plumb line and bob. Next, make two similar V-grooved sight vanes of hardboard or other rigid material. In one vane make a hole below the V of $\frac{1}{10}$ in (2·5 mm) diameter and in the other a hole of the smallest possible diameter whose centre is the same distance below the V as the other. Finally, fit the vanes to the ends of the batten so that lines joining the V-grooves and the centres of the holes are each parallel to the straight face of the batten.

To measure the altitude of a star, hold the batten steady and sight the star in line with the bottoms of the V-grooves. It may be found better to rest one end of the batten on something solid. Wait until the plumb line stops swinging and get a companion to note the altitude reading where the plumb line cuts the scale. If you are working alone, press the line to the scale with finger and thumb and take it to the light to read. The procedure for observing the sun is similar except that the sight holes are used, the smaller hole being nearer the eye, but before carrying out observations attach several layers of old, exposed photographic film over the larger sight hole to protect the eyes from direct sunlight. Whilst we are on the subject of observing the sun, never, on any account, look at the sun through any form of optical aid, otherwise permanent damage to the eyes may result—if there's one thing guaranteed to hamper a budding astronomer, it's blindness!

3. FINDING LATITUDE

An observer in the northern hemisphere may obtain his approximate latitude simply by measuring the altitude of the Pole Star. Since the Pole Star traces out a small circular path around the celestial pole during the course of a day, a better result could be achieved by making two observations, each 12 hours apart, and taking as the latitude the average of the two altitudes obtained, a method which is quite feasible during the long nights of winter. An alternative method is to measure the altitude of the sun, or a star, at the instant it is on the observer's meridian, when the altitude is greatest. If, as is more likely, the times of transit are not available, the next best thing is to erect vertical poles along the meridian and take the altitude exactly at the instant the sun or star is in the line of sight along the poles. If two persons are working together then, starting just before meridian passage, one could keep the object under continuous observation while the other keeps a close watch on the plumb line over the altitude scale until the maximum reading is reached. This method of waiting for the maximum altitude is frequently used by practical navigators. Once the meridian altitude has been obtained the procedure for calculating the latitude is as follows:

(a) Subtract the altitude from 90° to get the zenith distance.
(b) Name the zenith distance opposite to the azimuth. If the body observed was due south on the meridian, name the zenith distance north, and vice versa.
(c) Apply the declination, appropriately named north or south, to the zenith distance and observe the rule:—

like names—add
unlike names—subtract

The result is the latitude, which always takes the name of the larger quantity, not that any observer should be in any doubt as to which hemisphere he is situated. The following examples will make the method clear.

Example 1.

On July 20 the altitude of the sun when on the meridian due south of an observer was 57°·3. From the Sun Table at the end of this chapter, the declination of the sun on that date is found to be 20°·7 N.

	90°·0		
altitude	57°·3	south	
zenith distance	32°·7	north	same names—add
declination	20°·7	north	
latitude	53°·4	north	

An alternative method of finding the sun's meridian altitude is to use a shadow stick mounted vertically upon a perfectly flat baseboard of sufficient size to contain the shadow. The maximum altitude required for the latitude calculation occurs when the shadow is shortest, but in order to detect this the shadow must be checked continuously for some time before and after noon. The altitude can be obtained by scale drawing or by the formula:

$$\text{cotangent altitude} = \frac{\text{length of shadow}}{\text{length of stick}}$$

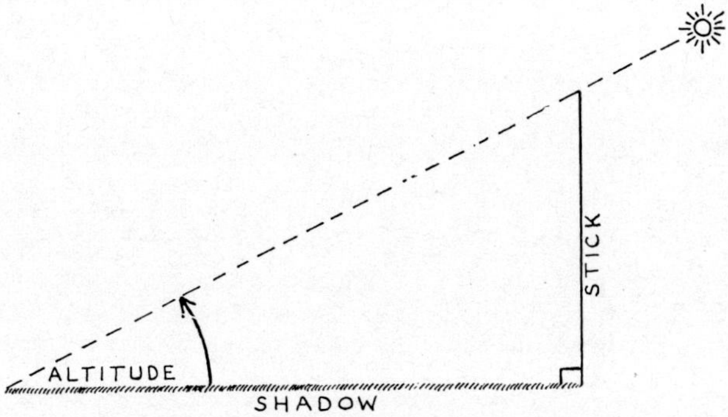

Fig. 1.16
The Sun's Altitude by Shadow Stick.

Suppose a stick three feet in height cast a shadow 1 ft 11 in in length. What is the altitude of the sun?

$$\text{cotangent altitude} = \frac{23}{36}$$

$$= 0{\cdot}6389$$

$$\text{altitude} = 57^\circ{\cdot}4$$

The same result would have been obtained with a stick one metre in height casting a noon shadow 64 centimetres in length. Once the altitude is known, the latitude is calculated as shown in example 1.

Example 2.

The star Rigel was observed on the meridian due south of an observer at an altitude of 28°·2. From the Star Table in Chapter 2 the declination of Rigel is found to be 8°·3 S.

	90°·0		
altitude	28°·2	south	
zenith distance	61°·8	north	different names—subtract
declination	8°·3	south	
latitude	53°·5	north	

A circumpolar star may also be observed at its lower transit, but in such a case the least altitude has to be waited for and the calculation is different to that for an upper transit observation. The declination is subtracted from 90° to obtain the polar distance, which is then added to the altitude to give the latitude.

Example 3.

The star Dubhe was observed at lower transit at an altitude of 25°·3. From the Star Table the declination of Dubhe is found to be 61°·9 N.

	90°·0	
declination	61°·9	
polar distance	28°·1	always add
altitude	25°·3	
latitude	53°·4 north	

Using the methods described, and taking the mean of several observations it should be possible to determine the latitude with an accuracy of between 0°·2 and 0°·4, provided the apparatus for measuring altitudes is constructed and used carefully.

Refraction, or the bending of the light from celestial objects as the rays pass through the earth's atmosphere could be a source of error. It is greatest near the horizon where it amounts to over ½°, as evidenced by the flattening of the sun's disc around sunrise and sunset, but decreases with increase of altitude until it becomes zero at the zenith. However, as far as we are concerned, if observations are confined to objects having altitudes of over 20° the effect of refraction is small enough to be ignored.

4. ANNUAL MOTION OF THE SUN

The apparent path of the sun around the celestial sphere during the course of a year (the ecliptic) may be shown by plotting the changing values of its Right Ascension and declination on a chart

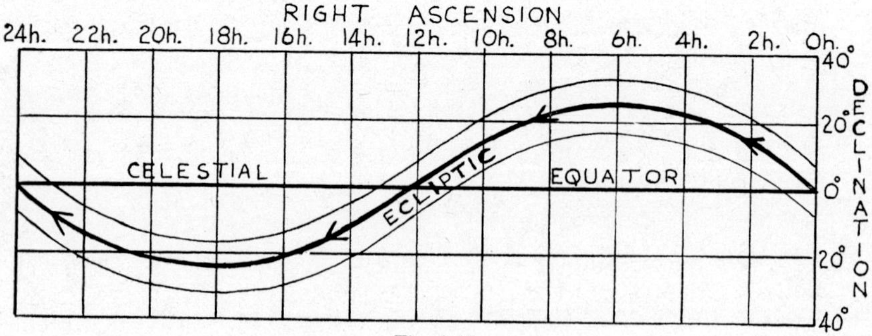

Fig. 1.17
The Sun's Apparent Path. The Zodiac.

employing a suitable system of reference lines. On a sheet of drawing paper construct a rectangular outline 12 in in length and 4 in in depth, (or 24 cm by 8 cm), inserting a network of faint lines $\frac{1}{2}$ in apart (1 cm), vertically to represent lines of Right Ascension and horizontally to represent parallels of declination. The Right Ascension scale is numbered progressively from right to left (eastwards), at hourly intervals, from 0 to 24 hours, each $\frac{1}{10}$ in (2 mm) being equivalent to 12 minutes.

The central horizontal line is the celestial equator and the declination scale is numbered at 10° intervals from 0° at the equator to 40° north (above) and south (below) of the equator, so that each $\frac{1}{10}$ inch (2 mm) is equivalent to 2°. The resulting grid is somewhat similar to that of the Mercator projection, often used for maps of the world, except that we have taken the liberty of spacing the parallels of declination equally, whereas on the Mercator map the parallels of latitude become further apart with increasing distance from the equator. Using the values of Right Ascension and declination given in the Sun Table, successive positions of the sun may be plotted on the grid and joined up to provide a smooth curve representing the ecliptic, on which the equinoxes and solstices should be indicated. The curve may not look like a great circle due to the unavoidable distortion, common to all maps, when a spherical surface is represented on a flat surface, but if the paper is rolled up into a vertical cylinder the great circle nature of the ecliptic will become readily apparent. Curves drawn parallel to and 8° on either side of the ecliptic gives the zodiacal belt which may be divided into 12 equal zones and indicated with the Signs of the Zodiac, commencing from 0 hours R.A. with the sign of Aries. As an added interest the position of some stars taken from the Star Table may be included on the map. This will reveal the discrepancy between the constellations and the zodiacal signs bearing the same names due to the precession of the equinoxes. It will also show the eastward movement of the sun with respect to the stars or, looked at in another way, the progressive westward drift of the stars with respect to the sun, a question to which we shall return later.

5. PHOTOGRAPHY—DAILY MOTION OF THE STARS

The apparent diurnal motion of the stars around the celestial sphere may be clearly demonstrated by photography, and for this purpose the most simple type of camera, using black and white film, will suffice. The most favourable conditions for taking photographs occur on calm, clear, moonless nights when the stars appear to shine brightly against an inky black background sky. It is essential to select a position well clear of house and street lighting, otherwise the film will very quickly fog over when the shutter is opened. It is also necessary to have some means of rigidly supporting the camera. A photograph similar to the one shown in Plate I may be obtained by fixing the camera in the direction of the Pole Star, opening the shutter for 20 minutes, and allowing the star images to trail over the film. The centre of the mass of star trails is the celestial pole, about which the stars are apparently rotating, the bright spot close to the centre being the Pole Star itself. Longer trails may be obtained by keeping the shutter open for a greater length of time. Exposures taken with the camera fixed in other directions, such as the east or west horizons, or elevated in the meridian towards the celestial equator (co-latitude), will yield further interesting and instructive photographs.

ANSWERS TO REVIEW QUESTIONS

2. Lat. 43° 42′ N, long. 79° 25′ W.
4. (a) 045°, 30° (b) 200°, 60° (c) 300°, 45°.
5. Celestial poles, E and W.
6. N, S.
7. 51° 30′, 38° 55′, 0° approx.
8. 50°.
9. Right Ascension, declination, ecliptic.

TABLE 1.1

THE SUN'S RIGHT ASCENSION AND DECLINATION
THE EQUATION OF TIME

DATE	R.A.		DECL.	E.o.T.
	hrs.	mins.	degrees	mins.
Jan. 1	18	45	—23·0	+ 3·5
6	19	07	—22·5	+ 5·8
11	19	29	—21·8	+ 7·9
16	19	51	—21·0	+ 9·7
21	20	12	—19·9	+ 11·3
26	20	33	—18·8	+ 12·6
31	20	53	—17·4	+ 13·5
Feb. 5	21	14	—16·0	+ 14·1
10	21	34	—14·4	+ 14·3
15	21	53	—12·7	+ 14·2
20	22	13	—11·0	+ 13·8
25	22	32	— 9·2	+ 13·2
Mar. 2	22	51	— 7·3	+ 12·3
7	23	10	— 5·3	+ 11·2
12	23	28	— 3·4	+ 9·9
17	23	46	— 1·0	+ 8·5
22	00	04	+ 0·6	+ 7·0
27	00	22	+ 2·5	+ 5·5
Apr. 1	00	40	+ 4·5	+ 4·0
6	00	59	+ 6·4	+ 2·5
11	01	17	+ 8·3	+ 1·1
16	01	36	+ 10·1	— 0·1
21	01	54	+ 11·8	— 1·3
26	02	13	+ 13·5	— 2·2
May 1	02	32	+ 15·0	— 2·9
6	02	51	+ 16·5	— 3·4
11	03	11	+ 17·8	— 3·7
16	03	30	+ 19·1	— 3·7
21	03	50	+ 20·2	— 3·5
26	04	10	+ 21·1	— 3·1
31	04	31	+ 21·9	— 2·5
June 5	04	51	+ 22·5	— 1·7
10	05	12	+ 23·0	— 0·8
15	05	32	+ 23·3	+ 0·3
20	05	53	+ 23·4	+ 1·4
25	06	14	+ 23·4	+ 2·5
30	06	35	+ 23·2	+ 3·5

TABLE 1.1 (continued)

DATE	R.A.		DECL.	E.o.T.
	hrs.	mins.	degrees	mins.
July 5	06	56	+ 22·8	+ 4·4
10	07	16	+ 22·3	+ 5·2
15	07	36	+ 21·6	+ 5·8
20	07	56	+ 20·7	+ 6·3
25	08	16	+ 19·7	+ 6·4
30	08	36	+ 18·6	+ 6·4
Aug. 4	08	55	+ 17·3	+ 6·1
9	09	14	+ 15·9	+ 5·5
14	09	33	+ 14·4	+ 4·7
19	09	52	+ 12·8	+ 3·7
24	10	10	+ 11·2	+ 2·4
29	10	29	+ 9·4	+ 1·0
Sept. 3	10	47	+ 7·6	— 0·6
8	11	05	+ 5·8	— 2·2
13	11	23	+ 3·9	— 4·0
18	11	41	+ 1·9	— 5·7
23	11	59	— 0·1	— 7·5
28	12	17	— 2·0	— 9·2
Oct. 3	12	35	— 3·9	— 10·9
8	12	53	— 5·8	— 12·3
13	13	12	— 7·7	— 13·7
18	13	30	— 9·6	— 14·7
23	13	49	— 11·3	— 15·6
28	14	08	— 13·1	— 16·2
Nov. 2	14	28	— 14·7	— 16·4
7	14	48	— 16·2	— 16·3
12	15	08	— 17·7	— 15·9
17	15	28	— 18·9	— 15·1
22	15	49	— 20·1	— 13·9
27	16	10	— 21·1	— 12·5
Dec. 2	16	32	— 21·9	— 10·7
7	16	54	— 22·6	— 8·7
12	17	16	— 23·1	— 6·4
17	17	38	— 23·4	— 4·0
22	18	00	— 23·4	— 1·5
27	18	22	— 23·3	+ 0·9

CHAPTER 2

STARS IN THE SKY

On a clear night, away from the glare of town lights, the night sky presents one of the most wonderful sights in nature. At times when the moon is absent from the heavens, the stars are revealed in all their splendour, like jewels of varying colour and brightness set upon black velvet. The stars are really suns, although in size and hotness they can be quite different from each other and from our own sun. It is because they are so far away from us that we cannot feel their heat, and see them only as points of light. Under favourable conditions up to 3000 stars are visible to the naked eye at any time.

THE CONSTELLATIONS

An observer even casually surveying the star-studded sky cannot fail to notice that many stars fall into well defined groups. Such patterns, called *constellations*, were also recognised by ancient astronomers who, allowing their imaginations to run riot, associated the star groups with figures of legendary or mythological beings and named them accordingly. When Ptolemy published his Almagest in about A.D. 150, nearly 50 constellations were referred to, many of their names like Orion, Hercules, Perseus and Cassiopeia, still being with us today. The early constellations did not cover the whole sky and their irregular boundaries were the result of their being made to fit the figures the stars were supposed to represent. Following the invention of the telescope, the subsequent establishment of observatories in the southern hemisphere allowed astronomers to map the whole sky and new constellations in the southern celestial hemisphere, previously unobserved, were given more modern names such as Telescopium and Microscopium. Today 88 constellations are recognised, whose boundaries are generally defined by lines of Right Ascension and parallels of declination. Early star charts and star globes were cluttered up with detailed drawings of the legendary figures, but these are omitted nowadays in favour of greater clarity. The constellations are useful inasmuch as they provide ready reference areas in the sky, particularly for nearer celestial bodies such as the moon and planets which move against the background of the stars. For

Table 2.1

Northern Constellations

Andromeda:	Andromeda.	Hercules:	Hercules.
Aquila:	Eagle.	Lacerta:	Lizard.
Auriga:	Charioteer.	Leo Minor:	Smaller Lion
Bootes:	Herdsman.	Lynx:	Lynx.
Camelopardus:	Giraffe.	Lyra:	Lyre.
Canes Venatici:	Hunting Dogs.	Ophiuchus:	Serpent Holder.
Cassiopeia.	Cassiopeia.	Pegasus:	Pegasus.
Cepheus:	Cepheus.	Perseus:	Perseus.
Coma Berenices:	Berenices Hair.	Sagitta:	Arrow.
Corona Borealis:	Northern Crown	Serpens:	Serpent.
Cygnus:	Swan.	Triangulum:	Triangle.
Delphinus:	Dolphin.	Ursa Major:	Great Bear.
Draco:	Dragon.	Ursa Minor:	Smaller Bear.
Equuleus:	Little Horse.	Vulpecula:	Fox.

Central (Zodiacal) Constellations

Aquarius:	Water Carrier.	Libra:	Scales.
Aries:	Ram.	Pisces:	Fishes.
Cancer:	Crab.	Sagittarius:	Archer.
Capricornus:	Sea Goat.	Scorpio:	Scorpion.
Gemini:	Twins.	Taurus:	Bull.
Leo:	Lion.	Virgo:	Virgin.

Southern Constellations

Antlia:	Air Pump.	Lupus:	Wolf.
Apus:	Bird of Paradise.	Mensa:	Table Mountain.
Ara:	Altar.	Microscopium:	Microscope.
Caelum:	Graving Tool.	Monoceros:	Unicorn.
Canis Major:	Larger Dog.	Musca:	Fly.
Canis Minor:	Smaller Dog.	Norma:	Level.
Carina:	Keel.	Octans:	Octant.
Centaurus:	Centaur.	Orion:	Orion.
Cetus:	Whale.	Pavo:	Peacock.
Chamaelon:	Chameleon.	Phoenix:	Phoenix.
Circinus:	Compasses.	Pictor:	Easel.
Columba:	Dove.	Pisces Austrinus:	Southern Fish.
Corona Australis:	Southern Crown.	Puppis:	Stern.
Corvus:	Crow.	Pyxis:	Mariners' Compass.
Crater:	Cup.	Reticulum:	Net.
Crux:	Cross.	Sculptor:	Sculptor's Apparatus.
Dorado:	Dorado.	Scutum:	Shield.
Eridanus.	River.	Sextans:	Sextant.
Fornax:	Furnace.	Telescopium:	Telescope.
Grus:	Crane.	Triangulum Australe:	Southern Triangle.
Horologium:	Clock.	Tucana:	Toucan.
Hydra:	Sea Serpent.	Vela:	Sails.
Hydrus:	Water Snake.	Volans:	Flying Fish.
Indus:	Indian.		
Lepus:	Hare.		

example, if Mars was said to be in Leo, an observer looking towards that constellation would readily locate the planet. Of course, the constellations are not like "pictures hanging on a wall", for the stars are at greatly varying distances from us and the patterns we see are particular to our position in space. To an observer on an imaginary planet of a distant star the constellations would appear entirely different from those we see.

DESIGNATIONS OF THE STARS

The names by which we know the brightest stars are mainly Greek or Arabic in origin—Sirius, Capella, Vega, Rigel, Aldebaran, Altair, etc. Over 350 years ago, the German astronomer Bayer introduced a system of designating the stars by letters of the Greek alphabet. Accordingly, the letter alpha is assigned to the brightest star in a constellation, beta to the second brightest, and so on. For example, Vega, the brightest star in the constellation Lyra, becomes known as Alpha Lyrae, and Algol, the second brightest star in Perseus, as Beta Persei. Upon the exhaustion of the Greek alphabet designations are continued in order of the Roman alphabet, a, b, c, d, etc. A numerical system devised by Flamsteed, the first British Astronomer Royal, was introduced in 1725 to cope with the ever increasing number of stars being discovered with the telescope, the stars of a constellation being numbered in increasing order of Right Ascension, 1, 2, 3, etc. Thus Arcturus, or Alpha Bootis, becomes 16 Bootis. Most maps of the

Table 2.2.
The Greek Alphabet
(Small Letters)

α alpha	ι iota	ρ rho
β beta	κ kappa	σ sigma
γ gamma	λ lambda	τ tau
δ delta	μ mu	υ upsilon
ε epsilon	ν nu	φ phi
ζ zeta	ξ xi	χ chi
η eta	ο omicron	ψ psi
θ theta	π pi	ω omega

naked-eye stars use a combination of the above mentioned methods. In more specialised catalogues, a star or other celestial object is designated by its number in that catalogue.

MAGNITUDES OF THE STARS

If two torchlights of unequal power are arranged at the same distance from an observer, the more powerful of the two will obviously appear brighter and the observer may well think it to be closer to him than the less powerful one. On the other hand, if the weaker light is brought closer, or the stronger one taken further away, it is possible to make the weaker light appear brighter to the observer. In other words, the apparent brightness of the lights is governed not only upon their intrinsic or real brightness, but also upon their distance. Similarly, an intrinsically faint star can appear brighter than an intrinsically bright star which happens to be much further away from us. Thus the brightness of the stars as they are seen in the sky provides no indication of their distance or real brightness—only their apparent brightness, a measure of which is called *apparent magnitude*.

The first system of classifying the stars according to their apparent brightness originated from Ptolemy, who called the 20 brightest stars 1st magnitude and then assigned magnitudes to the fainter stars in order of diminishing brightness, down to 6th magnitude for those just visible to the naked eye. This rough classification persisted until the early 19th century, when it was found that the intensity of the light we receive from a 1st magnitude star is about 100 times greater than that from a 6th magnitude star. If the figure 2·512 is multiplied five times, $2{\cdot}512^5$, the result is 100, which is the difference in brightness of two stars whose magnitude difference is 5. Therefore, a star of any given magnitude is 2·512 times as bright as one of the next lower magnitude. Unfortunately, some of the original 1st magnitude stars were much brighter than others, but the difficulty was overcome by taking magnitude 1 as the average brightness of these stars. This meant extending the scale to 0, and even into negative figures in order to accommodate the very bright stars. The scale was also extended upwards to take account of fainter telescopic stars, and the improved techniques of determining stellar magnitudes led to the introduction of decimal values. Under the present system:

A mag. —1 star is 2·512 times brighter than a star of mag. 0
,, mag. 0 ,, ,, 2·512 ,, ,, ,, ,, ,, ,, mag. 1
,, mag. 1 ,, ,, 2·512 ,, ,, ,, ,, ,, ,, mag. 2,
and so on.

The magnitude of the brightest star Sirius is —1·6, that of the planet Venus at its brightest —4·4, the full moon —12·6, and the sun —26·8. Through the giant 200-in (5·08-m) Hale telescope at Mount Palomar, USA, it is possible to see stars as faint as magnitude 20, although photography allows objects of magnitude 23 to be detected.

APPARENT MOTION OF THE STARS

Because the stars are so far away from us their positions on the celestial sphere, denoted by the co-ordinates of Right Ascension and declination, can be regarded as fixed. It is true that these co-ordinates show some slight change due to precession of the equinoxes and also because the stars have a true motion of their own in space, but the difference in their relative positions resulting from any real motion they have is so minute that in anyone's lifetime the constellations show no change. If the stars could be seen as they were, say 100 000 years ago some of the familiar constellations would appear quite different to what they do now. Although astronomers take into account the small changes in the positions of the stars which take place from one year to the next, we can afford to ignore them for our purposes without incurring any significant error.

Having stated that the stars can be considered as fixed points, it follows that their movement as seen in the sky is apparent rather than real. The diurnal motion they have due to the rotation of the earth on its axis, as seen from different latitudes has already been described in Chapter 1, and need not be pursued further. We have also seen that the revolution of the earth around the sun causes the sun to have an apparent annual motion to the east with respect to the stars. This amounts to one degree of arc, or four minutes of time each day. Since our time system is closely related to the sun, and not the stars, it appears to us that the stars move westwards with respect to the sun, by the same daily amount. This cannot be appreciated in the day time when the stars are not visible, but at night this slow westward drift becomes more noticeable as the stars rise and set four minutes earlier each night, causing different constellations to present themselves in a given part of the sky with the change of the seasons. For instance, the well known constellation of Orion dominates the night sky in winter, but it is invisible in the summer months when it is in the background of the sun. Now, four minutes per day adds up to two hours in 30 days, so that any star crosses an observer's meridian earlier by two hours each month. Considering mighty Orion once again, this

TABLE 2.3
THE BRIGHTEST STARS

DESIGNATION	POPULAR NAME	R. A.		DECL.	MAG.
		hrs.	mins.	degrees	
α Pisces Australis	Fomalhaut	22	56	—29·8	1·3
α Cygni	Deneb	20	40	+45·2	1·3
α Aquilae	Altair	19	49	+ 8·8	0·9
α Lyrae	Vega	18	36	+38·8	0·1
ε Sagittarii	Kaus Australis	18	22	—34·4	1·9
θ Scorpii	Theta Scorpii	17	45	—43·0	2·0
λ Scorpii	Shaula	17	32	—37·1	1·7
α Triang. Australis	Atria	16	45	—69·0	1·9
α Scorpii	Antares	16	28	—26·5	1·2
α Centauri	Rigil Kentaurus	14	38	—60·7	0·1
α Bootes	Arcturus	14	14	+19·3	0·2
β Centauri	Hadar	14	02	—60·2	0·9
η Ursa Majoris	Alkaid	13	46	+49·4	1·9
α Virginis	Spica	13	24	—11·0	1·2
ε Ursa Majoris	Alioth	12	53	+56·1	1·7
β Crucis	Mimosa	12	46	—59·5	1·5
γ Crucis	Gacrux	12	30	—56·9	1·6
α Crucis	Acrux	12	25	—62·9	1·1
α Ursa Majoris	Dubhe	11	02	+61·9	1·9
α Leonis	Regulus	10	07	+12·1	1·3
β Carinae	Miaplacidus	09	13	—69·6	1·8
δ Velorum	Delta Velorum	08	44	—54·6	2·0
ε Carinae	Avior	08	22	—59·6	1·7
γ Velorum	Gamma Velorum	08	09	—47·2	1·9
β Geminorum	Pollux	07	44	+28·1	1·2
α Canis Minoris	Procyon	07	38	+ 5·3	0·5
α Geminorum	Castor	07	33	+32·0	1·6
δ Canis Majoris	Wezen	07	07	—26·3	2·0
ε Canis Majoris	Adhara	06	57	—28·9	1·6
α Canis Majoris	Sirius	06	44	—16·7	—1·6
γ Geminorum	Alhena	06	36	+16·4	1·9
α Carinae	Canopus	06	23	—52·7	—0·9
β Canis Majoris	Mirzam	06	21	—17·9	2·0
α Orionis	Betelgeuse	05	53	+ 7·4	0·1 to 1·2
κ Orionis	Saiph	05	46	— 9·7	2·2
ζ Orionis	Alnitak	05	39	— 2·0	2·0
ε Orionis	Alnilam	05	35	— 1·2	1·7
ι Orionis	Iota Orionis	05	34	— 5·9	2·9
δ Orionis	Mintaka	05	30	— 0·3	2·5
γ Orionis	Bellatrix	05	24	+ 6·3	1·7
β Tauri	Elnath	05	24	+28·6	1·8
α Aurigae	Capella	05	14	+46·0	0·2
β Orionis	Rigel	05	13	— 8·3	0·3
α Tauri	Aldebaran	04	34	+16·5	1·1
α Persei	Mirfak	03	22	+49·8	1·9
α Eridani	Achernar	01	36	—57·4	0·6

constellation crosses the meridian at about 11 P.M. local time in December; 9 P.M. in January; 7 P.M. in February; and so on. Of course, the changing pattern of the night sky repeats itself every year because the westward march of the constellations is a simple result of the earth's annual revolution around the sun.

THE STAR MAPS

Maps 1 and 2 show the bright stars and the major constellations of the northern and southern celestial hemispheres respectively. The lines radiating from their centres are hour circles numbered with hours of Right Ascension from 0 to 24. The circles are parallels of declination at intervals of 30°, with 10° marks in between.

An observer in northern latitude is in a position to see all the stars of the northern celestial hemisphere, together with southern stars having declinations of less than 90° minus his latitude. For example, from latitude 50° N it is possible to see stars south of the celestial equator down to 40° S. Similarly, an observer in latitude 30° S will see all the stars of the southern celestial hemisphere in addition to those northern stars whose declinations do not exceed 60° N. With these examples in mind an observer in any latitude can by drawing a circle on the appropriate star map—the southern map for a northern observer—determine which stars of the opposite hemisphere it is possible for him to see, and also those which he will never see. Of course, stars less than 5° above the horizon are unlikely to be seen due to surface haze unless they are very bright.

Since a star is circumpolar if 90° minus declination is less than the latitude, a circle of radius equal to the number of degrees in the observer's latitude and centred on the elevated pole will enclose all the stars which are above the horizon on each night of the year. If the map is rotated until the date on the outer scale is at the top, then the lower edge of this circle represents the northern horizon of a northern observer and the southern horizon of a southern observer. It should be noted that a star attains the zenith only if its declination is equal to the observer's latitude.

Maps 1 and 2 can also be used to find the approximate time of the meridian passage of a star, as described later in the Project Section.

Maps 3 to 11 inclusive, are incuded in order to provide more positive identification of stars and other interesting objects which figure in the Project Sections of this and other Chapters.

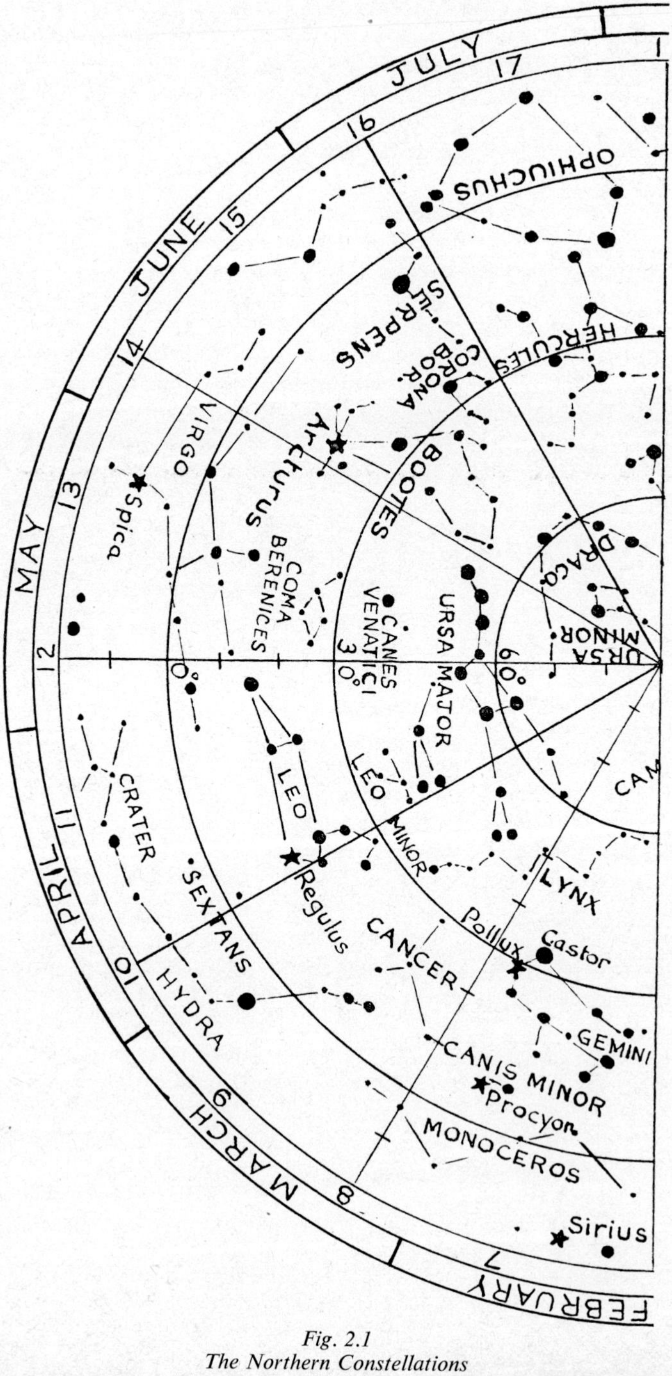

Fig. 2.1
The Northern Constellations

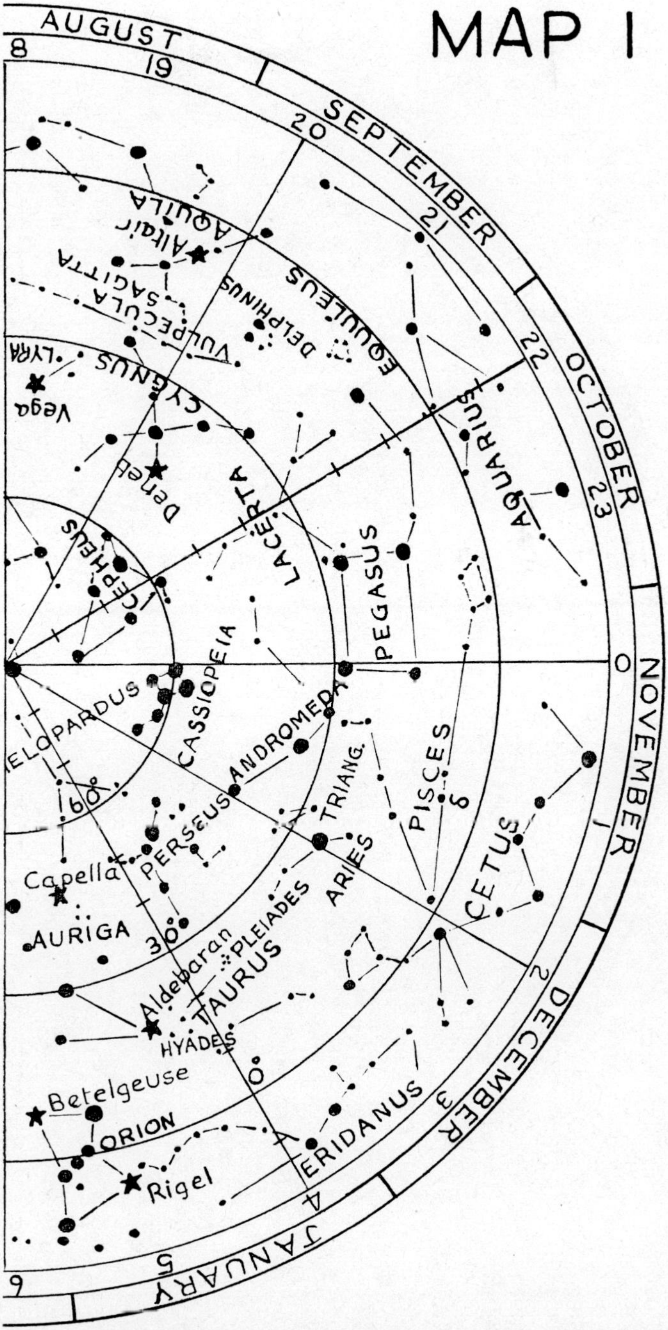
MAP I
AUGUST
SEPTEMBER
OCTOBER
NOVEMBER
DECEMBER
JANUARY
AQUILA
Altair
SAGITTA
VULPECULA
DELPHINUS
EQUULEUS
LYRA
Vega
CYGNUS
Deneb
CEPHEUS
LACERTA
PEGASUS
AQUARIUS
CASSIOPEIA
CAMELOPARDUS
ANDROMEDA
TRIANG.
PISCES
CETUS
PERSEUS
ARIES
Capella
AURIGA
PLEIADES
Aldebaran
TAURUS
HYADES
Betelgeuse
ORION
Rigel
ERIDANUS

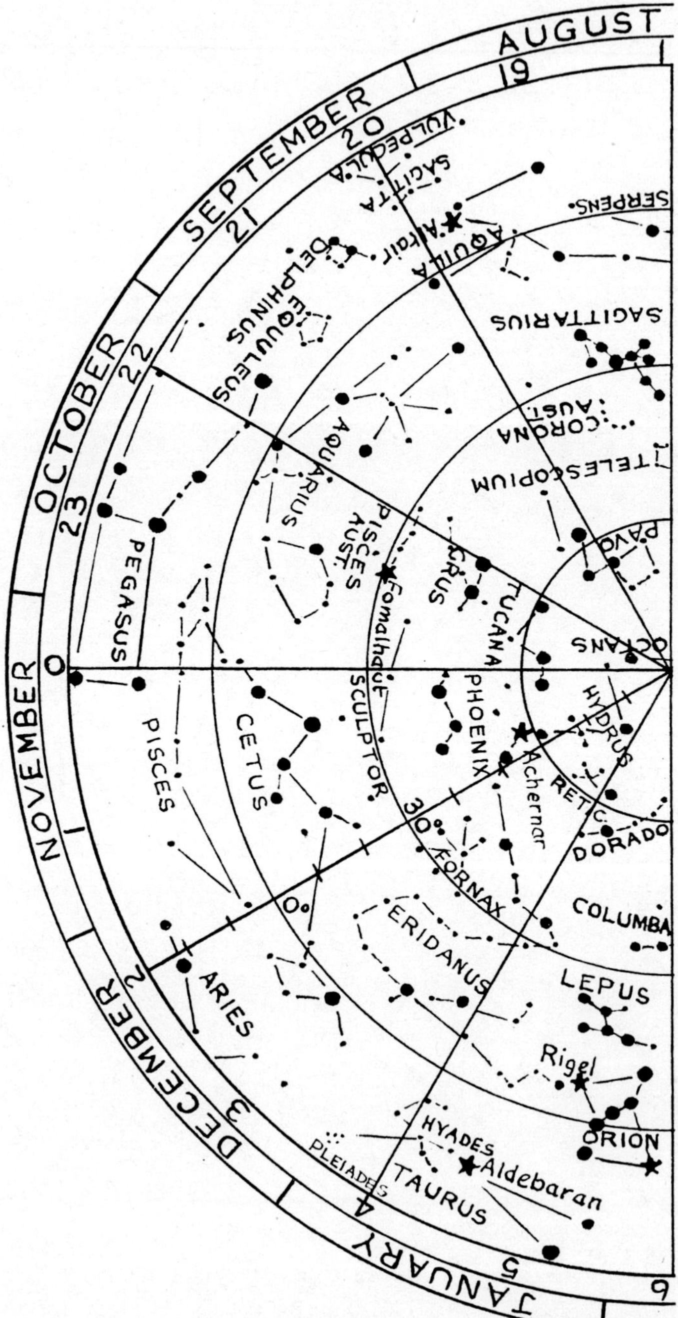

Fig. 2.2
The Southern Constellations.

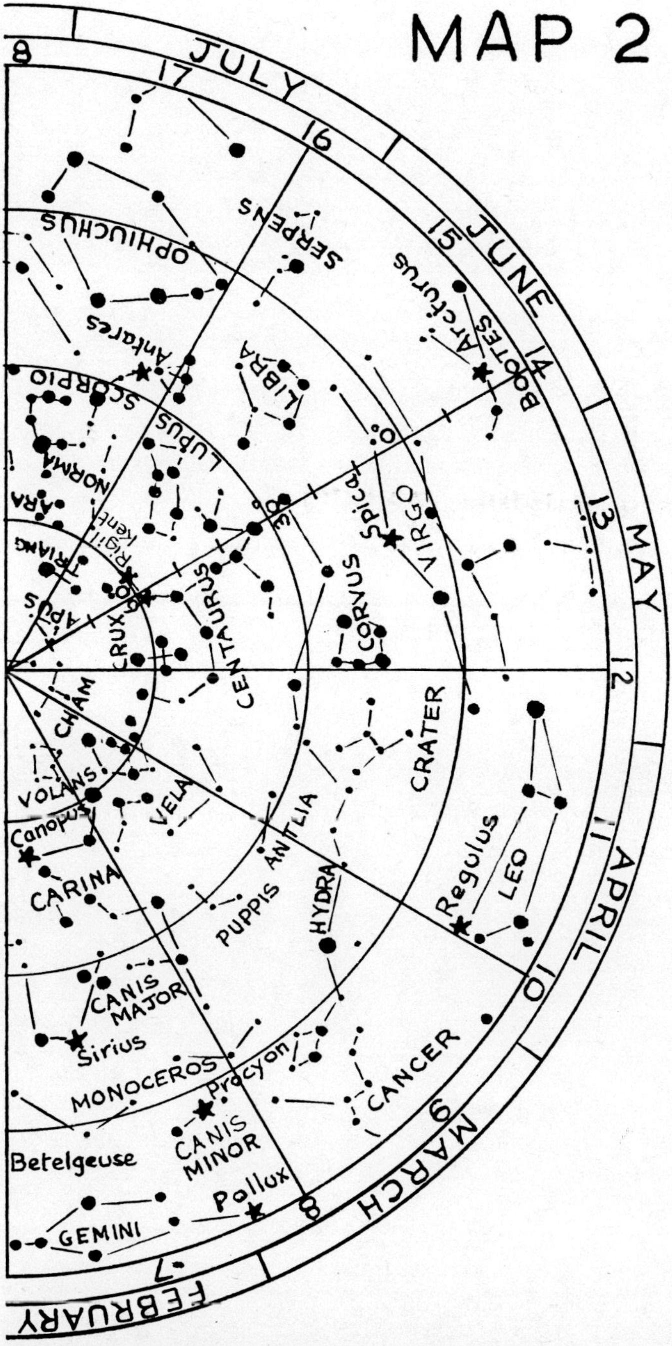
JULY
JUNE
MAY
APRIL
MARCH
FEBRUARY
8
17
16
15
14
13
12
11
10
9
8
7
30
SERPENS
OPHIUCHUS
Antares
SCORPIO
LIBRA
Arcturus
BOOTES
LUPUS
NORMA
ARA
Rigil Kent
TRIANG
APUS
CRUX
CENTAURUS
Spica
VIRGO
CORVUS
CRATER
CHAM
VOLANS
VELA
ANTLIA
Canopus
CARINA
PUPPIS
HYDRA
Regulus
LEO
CANIS MAJOR
Sirius
MONOCEROS
Procyon
CANCER
CANIS MINOR
Betelgeuse
Pollux
GEMINI

MAP 2

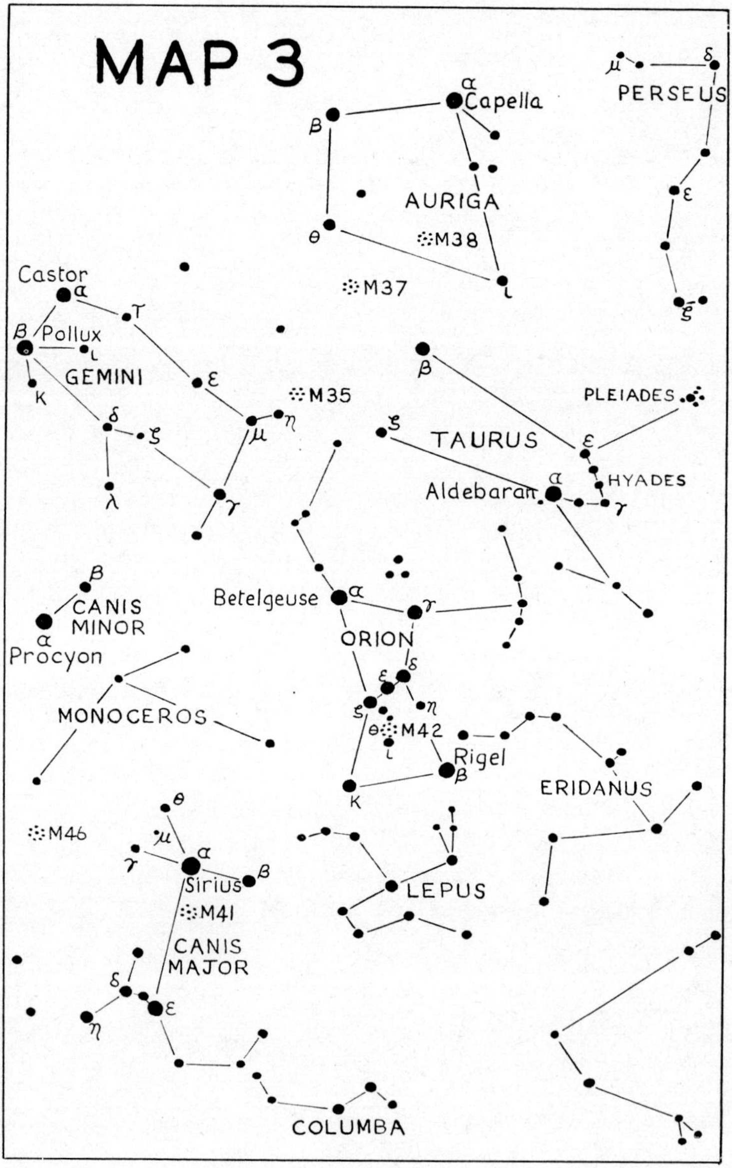
MAP 3
Capella
AURIGA
M38
M37
PERSEUS
Castor
Pollux
GEMINI
M35
PLEIADES
TAURUS
HYADES
Aldebaran
CANIS
MINOR
Procyon
Betelgeuse
ORION
MONOCEROS
M42
Rigel
ERIDANUS
M46
Sirius
M41
CANIS
MAJOR
LEPUS
COLUMBA

Fig. 2.3

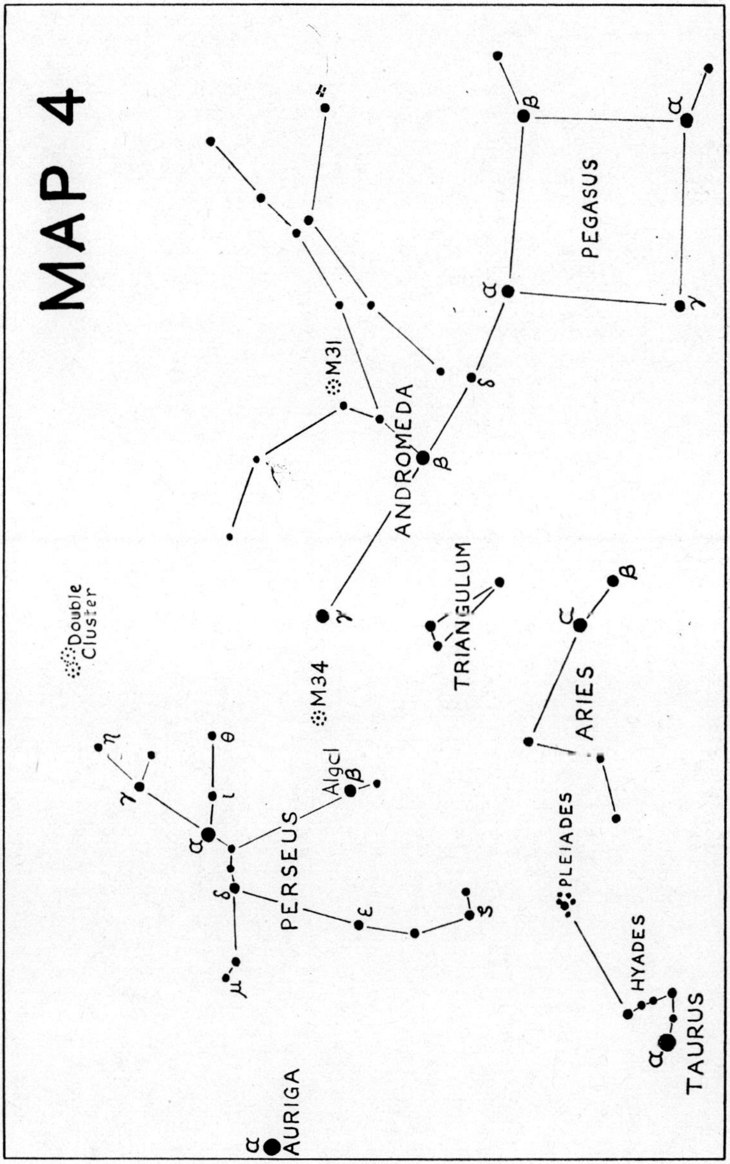
MAP 4
PEGASUS
ANDROMEDA
M31
Double Cluster
M34
TRIANGULUM
ARIES
PERSEUS
Algol
PLEIADES
HYADES
TAURUS
AURIGA

Fig. 2.4

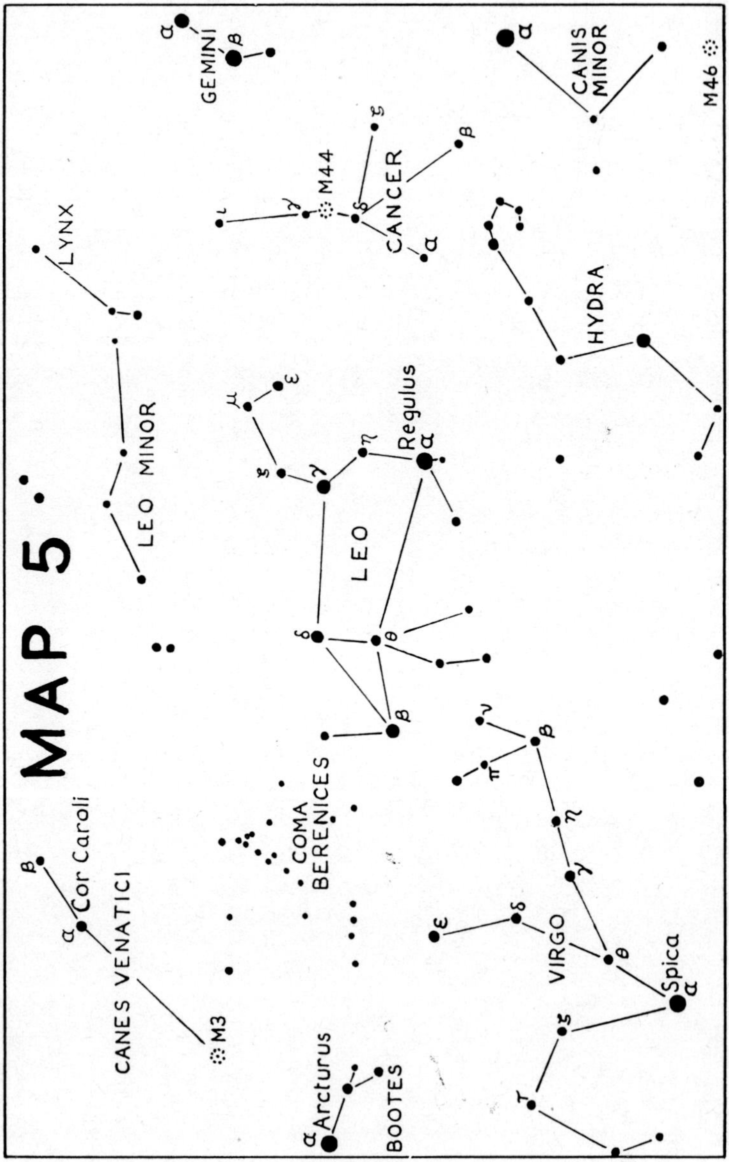
MAP 5
LYNX
GEMINI
M44
CANCER
CANIS MINOR
M46
LEO MINOR
LEO
Regulus
HYDRA
COMA BERENICES
Cor Caroli
CANES VENATICI
M3
Arcturus
BOOTES
VIRGO
Spica

Fig. 2.5

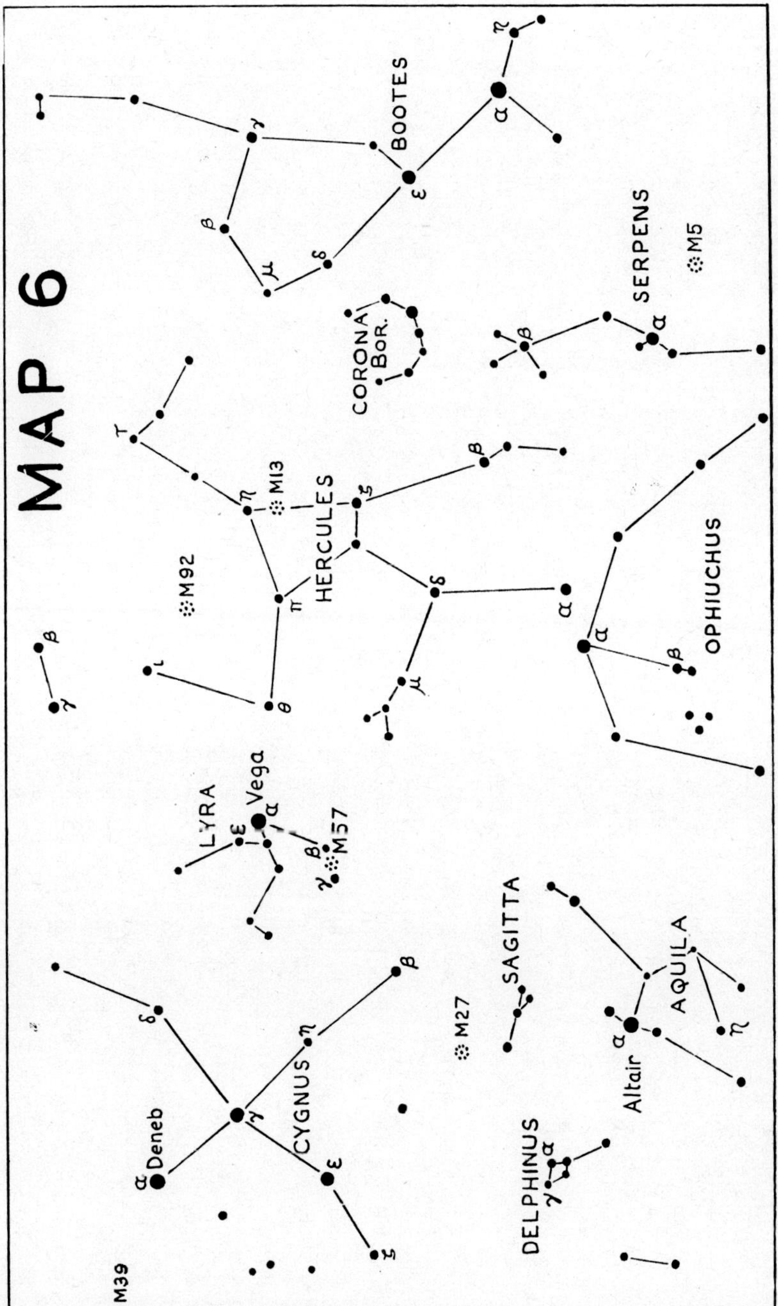
MAP 6
BOOTES
CORONA BoR.
SERPENS
M5
HERCULES
M13
M92
OPHIUCHUS
LYRA
Vega
M57
SAGITTA
AQUILA
Altair
M27
CYGNUS
Deneb
DELPHINUS
M39

Fig. 2.6

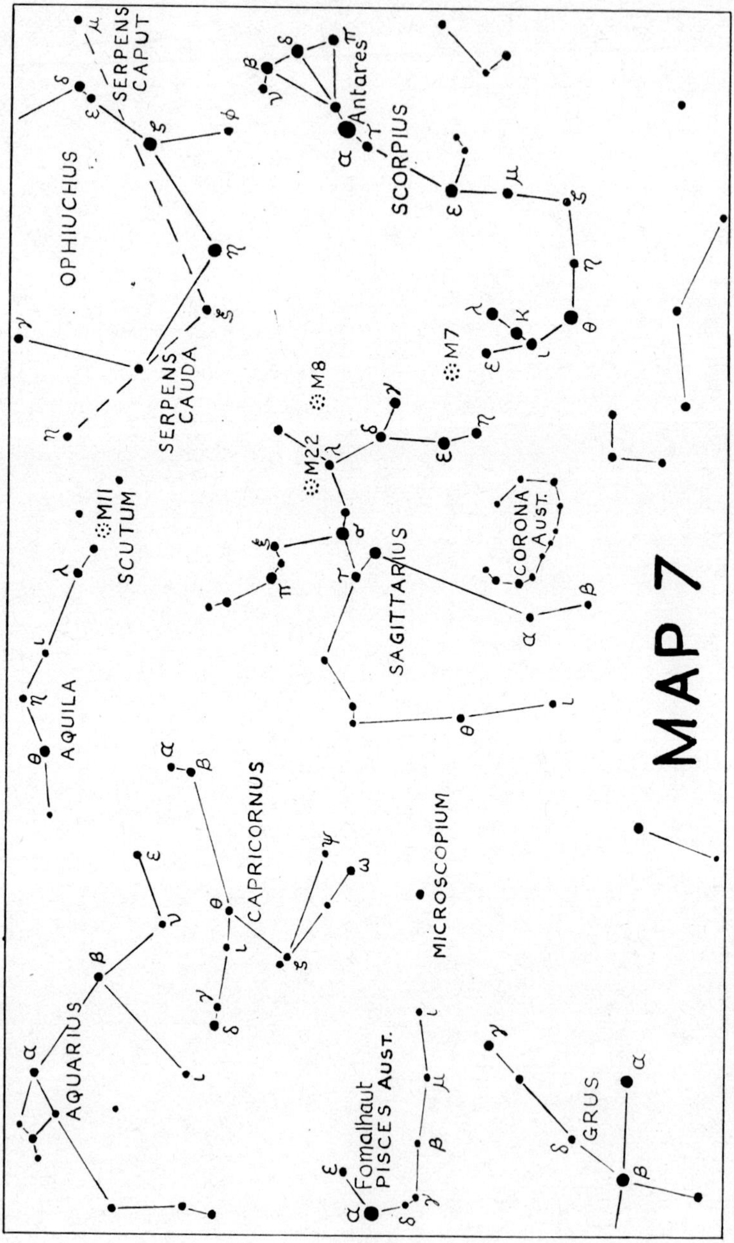
MAP 7
OPHIUCHUS
SERPENS CAPUT
SERPENS CAUDA
SCORPIUS
Antares
SCUTUM
M11
M22
M8
M7
AQUILA
SAGITTARIUS
CORONA AUST.
CAPRICORNUS
MICROSCOPIUM
AQUARIUS
Fomalhaut
PISCES AUST.
GRUS

Fig. 2.7

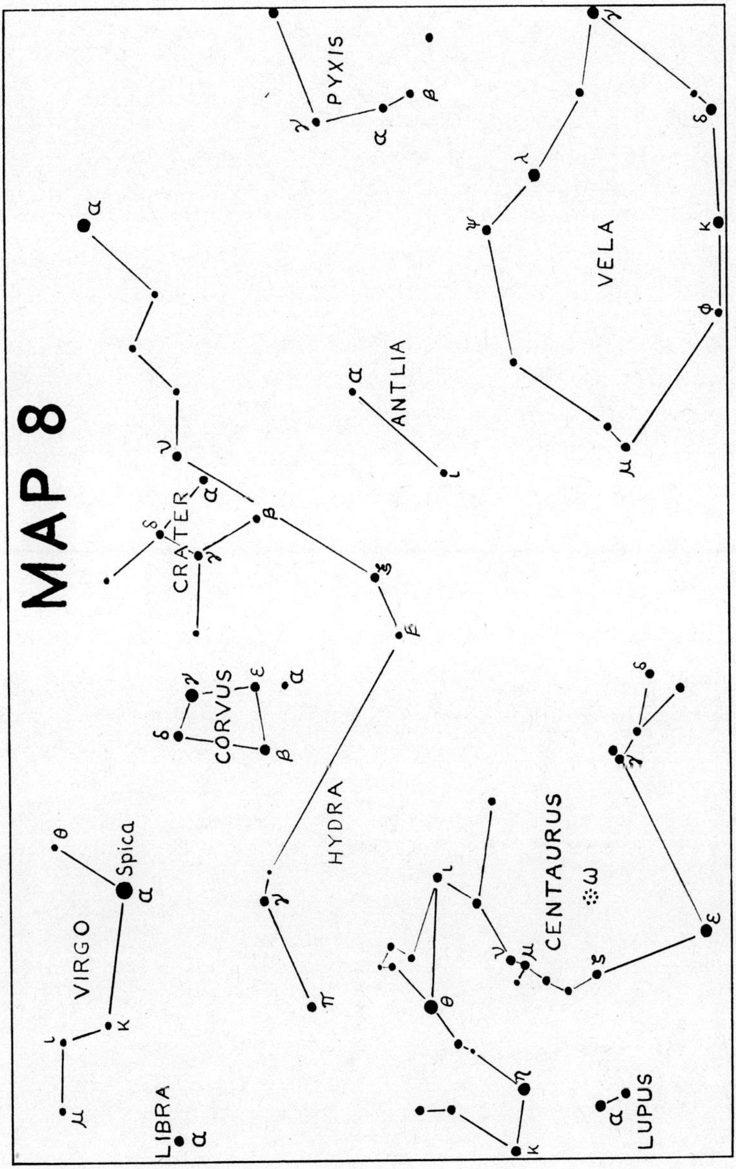
MAP 8
VIRGO
Spica
LIBRA
CRATER
CORVUS
HYDRA
ANTLIA
PYXIS
VELA
CENTAURUS
LUPUS

Fig. 2.8

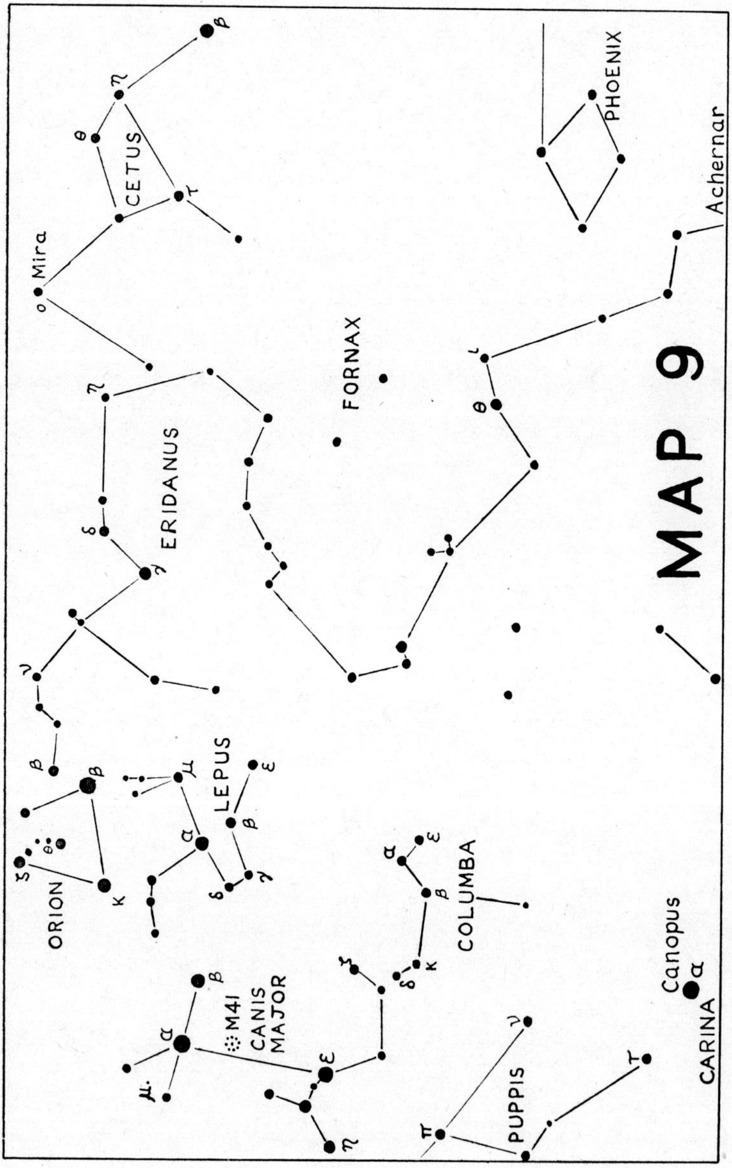
MAP 9
CETUS
Mira
PHOENIX
Achernar
FORNAX
ERIDANUS
LEPUS
ORION
M41
CANIS
MAJOR
COLUMBA
Canopus
CARINA
PUPPIS

Fig. 2.9

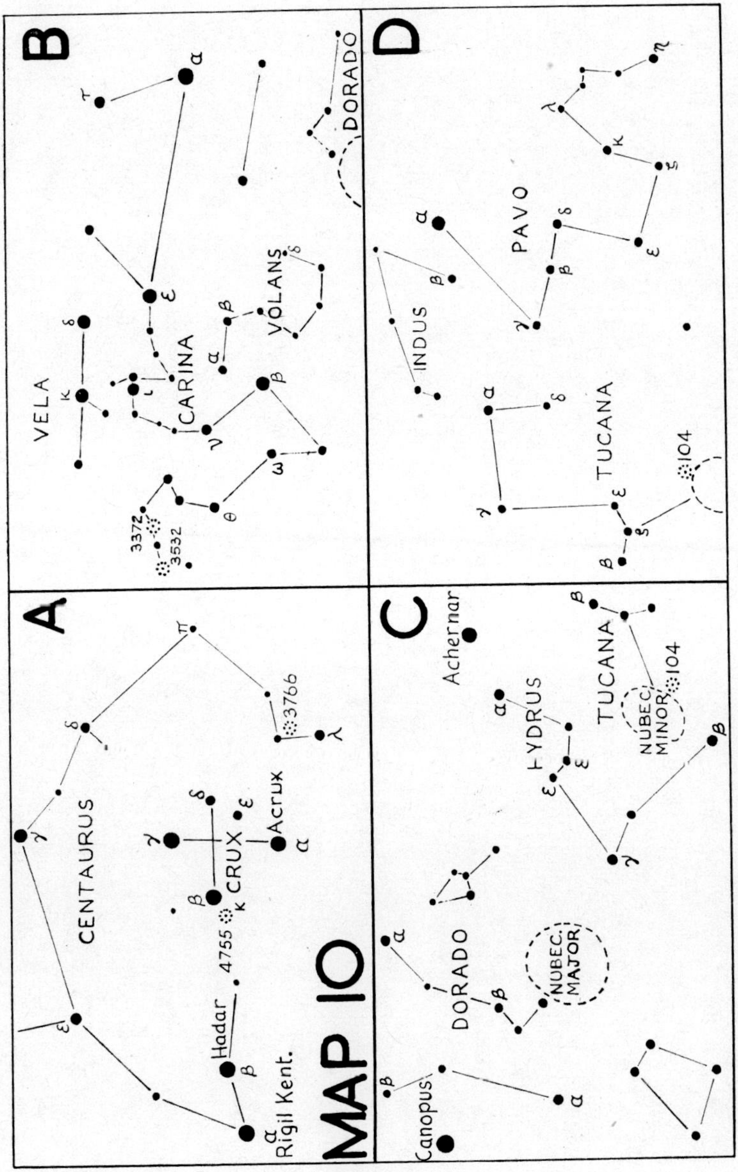
MAP 10
A
CENTAURUS
CRUX
Hadar
Rigil Kent.
Acrux
4755
3766
B
VELA
CARINA
VOLANS
DORADO
3372
3532
C
Canopus
DORADO
NUBEC. MAJOR
HYDRUS
TUCANA
NUBEC. MINOR
104
Achernar
D
INDUS
PAVO
TUCANA
104

Fig. 2.10

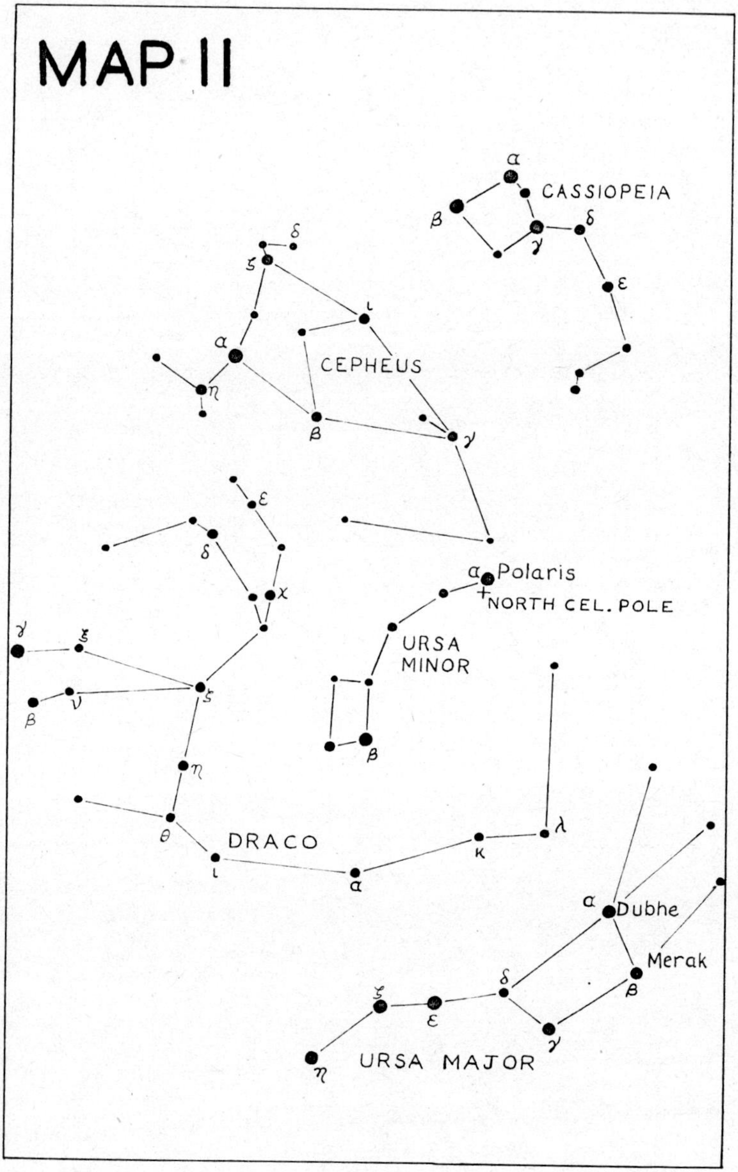
MAP II
CASSIOPEIA
CEPHEUS
Polaris
NORTH CEL. POLE
URSA MINOR
DRACO
Dubhe
Merak
URSA MAJOR

Fig. 2.11

REVIEW QUESTIONS

1. From where did many of the stars and constellations gain their names? Name six bright stars and constellations visible from your latitude.

2. What is apparent magnitude? Two stars have magnitudes —1 and +1 respectively. Which is the brighter of the two, and by how much is it brighter than the other?

3. After which animals do the following constellations take their names: Ursa Major, Cygnus, Leo, Aquilla and Taurus?

4. After which instruments do the following constellations take their names: Octans, Sextans, Horologium, Norma and Pyxis?

5. Name two constellations which are circumpolar from your latitude.

6. Which of the following stars never set, never rise, or rise and set at your latitude: Vega, Altair, Fomalhaut, Sirius, Canopus, Arcturus, Dubhe and Capella?

7. Do all the alpha stars in different constellations have the same apparent magnitude? The stars mentioned in Question 6 are all alpha stars: to which constellation does each belong?

8. With reference to Table 2.3, name the brightest star which nearly attains the zenith at the following places:
Edinburgh (56°·0 N), Washington D.C. (38°·9 N), Ottawa (45°·4 N), Durban (29°·9 S), Adelaide (34°·9 S), Auckland (36°·9 S).

PROJECTS AND RESEARCH

1. SURVEYING THE NIGHT SKY

There is only one way to learn the constellations, that is to go outdoors on a clear night and discover them for oneself; for this is one area of astronomy in which no amount of reading is a substitute for observation. However, some guidance may be appreciated and it is hoped that the notes given in this section will be found useful.

One of the best known constellations to northern observers is Ursa Major, the Great Bear, because seven bright stars contained within its boundaries form the familiar figure of The Plough (fig. 1.7). Our American cousins know The Plough as the Big Dipper; probably a more appropriate name since four stars mark the corners of the "bowl" and three form the "handle" of what appears to be a large spoon. It will be recalled that a star is circumpolar only if its declination subtracted from 90° is less than the latitude. Alkaid, whose declination, $49\frac{1}{2}°$ N, is the least of any stars of The Plough, is circumpolar in all latitudes higher than $40\frac{1}{2}°$ N; and since the other six stars have declinations greater than Alkaid, it follows that the whole of The Plough is circumpolar in these latitudes and remains above the horizon every night throughout the year. In lower latitudes it rises and sets, being visible for a shorter time the further south one goes. For instance, "cobbers" in northern Australia see The Plough barely raise itself above the northern horizon.

A line through the end stars of the bowl of The Plough, Merak and Dubhe, extended beyond Dubhe for about five times the apparent distance between these two stars, leads to a fairly bright isolated star of second magnitude. This is, of course, the Pole Star; an object which has already proved of great use to us in finding latitude and the direction of true north. Its proximity to the north celestial pole gives Polaris a fixed appearance in the sky, while other stars in its vicinity trace out their diurnal circles around it. The circumpolar motion of The Plough is particularly noticeable as it sweeps around the Pole like the hand of a gigantic clock; and although its motion is opposite to that of a conventional clock we shall see later how it can be used to tell the time. The Pole Star is the brightest star in the constellation Ursa Minor, the Little Bear; hence its proper name Alpha Ursa Minoris. The Little Bear is somewhat like The Plough in shape, only smaller; across the Atlantic it is called the Little Dipper. Polaris marks the tip of the handle and another fairly bright star, Kochab, is situated at one corner of its bowl. The other stars of this constellation are relatively faint, but can be seen quite well on a clear night.

Cassiopeia, another circumpolar constellation as seen from middle-northern latitudes, can be found by looking across the Pole Star from The Plough. It does not contain any stars of first magnitude, but since there are no particularly bright stars in its immediate neighbourhood its irregular W or M pattern stands out quite conspicuously.

With the aid of the star maps it should be possible to locate two other circumpolar constellations, Draco and Cepheus. Although they are not blessed with any very bright stars, they can be identified readily on a clear moonless night well away from bright lights.

Andromeda is found on the side of Cassiopeia opposite to the Pole Star and about the same distance away. It consists of several stars in the form of a long sweeping curve. At the end of the curve furthest away from Cassiopeia are four stars of between second and third magnitude, each of which is at the corner of a large square. This is the constellation Pegasus. Andromeda and Pegasus together look something like a large version of Ursa Minor. A line through the two right hand stars of the square of Pegasus, Scheat and Markab, leads directly to the Pole Star. In the early months of autumn, this line extended towards the southern horizon identifies lonely Fomalhaut in the constellation Pisces Austrinus. It is interesting to note that a line through the two left hand stars of the "square" almost coincides with the hour circle through the First Point of Aries (Vernal Equinox). Further to the left (east) of these two stars, below the tail end of Andromeda, lie two fairly bright isolated stars in the constellation Aries. Can you remember why the hour circle through the First Point of Aries does not pass through the constellation of Aries?

A continuation of the curve of Andromeda away from Pegasus leads to the fairly bright star Mirfak in the constellation Perseus. A further extension of this curve for about the same distance beyond Mirfak, arrives at Capella, the chief star in Auriga. Capella is very bright, and unmistakably yellowish in colour; its location can be confirmed via The Plough by taking a line from Megrez through Dubhe. On the opposite side of the pole to Capella, and almost equidistant from the Pole Star, lies the brilliant silvery blue star Vega, in Lyra. Both Vega and Capella are circumpolar to British observers. In late summer they see Vega almost overhead, with Capella low down near the northern horizon; in winter the situation is reversed, with Capella occupying the near zenithal position. Just to the west of Vega lies the tangled web of Hercules;

it contains no bright stars and its pattern may not be easily identified. On the other side of Vega, to the east, is the bright star Deneb which lies at the tail of Cygnus, the Swan; though it is sometimes known as the Northern Cross. A keen observer may detect the shape of this constellation, which does somewhat resemble the outline of a swan in flight heading towards the southern horizon. To the south of Vega and Deneb is another bright star, Altair, in the constellation Aquila; these three stars together form a large triangle, known as the "summer triangle" because it is conspicuous near the meridian at this time of the year. In midsummer, a bright reddish star hanging low over the southern horizon will be Antares, or Alpha Scorpii.

During late spring and early summer, the stars Arcturus and Spica, belonging to the constellations Bootes and Virgo respectively, can be located in order by following the curve of the handle of The Plough. Arcturus is one of the brightest stars in the whole sky and is easily recognised by its pale orange colour. Spica, however, is distinctly white, and being a southern star does not rise nearly so high in the sky as Arcturus.

During the winter months, the magnificent constellation of Orion, The Hunter, presents itself in all its splendour. The pattern of Orion is quite unmistakable, with four bright stars marking the corners of a quadrilateral measuring 16° by 10°; the orange-red Betelgeuse and the brilliant white Rigel, positioned at diagonally opposite corners, are both first magnitude stars. Running slantwise across the centre of the quadrilateral is the conspicuous "belt" of The Hunter, comprising three equidistantly spaced stars. The belt lies almost on the celestial equator so that Orion is seen equally well by observers in both hemispheres. The only difference is that to northern observers it dominates the sky to the south, whereas southern observers see it "upside down" to the north in the months of their summer.

Because it is so large and its stars so bright, Orion provides many useful signposts to other constellations. A line through the belt extended towards the south leads directly to Canis Major and Sirius, the Dog-Star, whose superior brilliance outshines all other stars in the sky. The line extended in the opposite direction leads to the northern constellation Taurus, with its major star Aldebaran situated at the end of one arm of a noticeable V-shaped combination of stars called the Hyades. Beyond the Hyades lies a compact cluster of stars resembling the shape of a small saucepan; its astronomical name is the Pleiades, though it is commonly known as the Seven Sisters because of the number of stars of the cluster seen with the naked eye by persons having normal sight.

The two bright stars Castor and Pollux in the constellation Gemini, The Twins, can be located by extending a line from Rigel through Betelgeuse and beyond it for about twice the distance between them.

Procyon, a first magnitude star in Canis Minor, can be found on a line through the two northern stars of Orion, Bellatrix and Betelgeuse; confirmation of identification lies in the fact that Procyon, Betelgeuse and Sirius form an equilateral triangle. A continuation of this line beyond Procyon for about the same distance again, passes through Leo whose chief star, Regulus, marks the dot in a group of stars arranged in the form of a large question mark, otherwise known as The Sickle.

Observers in the southern Australia, New Zealand, and the Cape Province can see several bright south circumpolar stars, denied to northern observers. These include: Canopus, surpassed in brilliance only by Sirius; Achernar, the only star of note in the long, winding constellation Eridanus; Alpha and Beta Centauri, two first magnitude stars very close together; and Acrux, the brightest star in Crux, The Southern Cross.

Of course, many constellations have been omitted from this brief survey of the night sky, but these may be found with the aid of the star maps, using as stepping stones those bright stars and constellations which have been mentioned.

2. STARS ON THE MERIDIAN

The altitude of a star, or any heavenly body for that matter, is greatest when it is on the observer's meridian. If both the latitude of the observer and the declination of the star are known, the altitude can be obtained by means of a simple calculation. The declination and the latitude, appropriately named north or south, are applied to each other in accordance with the rule:

different names—add
same names—subtract.

This provides the zenith distance, which is subtracted from 90° to give the altitude. Should the zenith distance be 90° or more, the star remains permanently below the observer's horizon. The examples show the simplicity of the method.

Example 1.

Can the star Canopus, decl. 52°·7 S, ever be seen from Plymouth, England, lat. 50°·4 N?

latitude	50°·4 north	different names—add
declination	52°·7 south	
zenith distance	103°·1	

The zenith distance exceeds 90°, thus Canopus will never be visible from Plymouth.

Example 2.

What is the altitude of the star Fomalhaut, decl. 29°·8 S, at upper transit of the meridian at New York, lat. 40°·7 N?

latitude	40°·7 north	different names —add
declination	29°·8 south	
zenith distance	70°·5	
	90°·0	
altitude	19°·5	

Example 3.

What will be the maximum altitude attained by the star Vega, decl. 38°·8 N, at London, lat. 51°·5 N?

latitude	51°·5 north	same names—subtract
declination	38°·8 north	
zenith distance	12°·7	
	90°·0	
altitude	77°·3	

Example 4.

Calculate the maximum altitude attained by the star Alpha Centauri, decl. 60°·7 S, as observed from Sydney, Australia, lat. 33°·9 S.

latitude	33°·9 south	same names—subtract
declination	60°·7 south	
zenith distance	26°·8	
	90°·0	
altitude	63°·2	

The minimum altitude of a circumpolar star occurs at lower transit, about 12 hours before or after upper transit of the meridian, and is calculated as follows:
subtract the declination from 90° to obtain the polar distance, which is then subtracted from the latitude to give the altitude.

Example 5.

Find the maximum and minimum altitudes of the star Dubhe, decl. 61°·9 N, at Newcastle, England, lat. 55°·0 N.

latitude	55°·0 north	same names—subtract.
declination	61°·9 north	
zenith distance	6°·9	
	90°·0	
maximum altitude	83°·1	

	90°·0	
declination	61°·9	
polar distance	28°·1	always subtract
latitude	55°·0	
minimum altitude	26°·9	

The approximate time of meridian passage of a star can be ascertained from the Maps 1 and 2. These are designed so that a radial line (hour circle) drawn from the pole through a given star, cuts the outer scale on the date on which that star crosses the meridian at 9 P.M. local time (10 P.M. British Standard Time). For other dates, one simply has to bear in mind that meridian passage occurs 4 minutes earlier each day after the 9 P.M. date, and 4 minutes later each day prior to this date. Remember, 4 minutes each day adds up to 2 hours every 30 days.

Example 6.

Find the date on which the star Rigel is on the meridian at 9 P.M. local time, and thence the dates on which it will transit the meridian at 7 P.M. and 10 P.M. respectively.

A careful examination of maps 1 or 2 shows that Rigel is on the meridian at 9 P.M. on January 24th.

Now 7 P.M. is 2 hours earlier than 9 P.M., therefore, Rigel crosses the meridian 30 days later than January 24th,

i.e. on February 23rd.

Similarly, 10 P.M. is 1 hour later than 9 P.M., therefore, Rigel crosses the meridian 15 days earlier than January 24th,

i.e. on January 9th.

Thus we have the means of finding both the altitude and the time of a star's meridian passage. The next step is to put theory into practice. Using the star maps as a guide, compile a list of the brightest stars visible from your locality. Calculate their maximum altitudes and forecast the dates on which they cross the meridian at some convenient evening hour. Then, as each star becomes visible check the accuracy of your predictions by observation.

3. A CIRCUMPOLAR STAR MAP

The making of such a map will help you to become more familiar with the area of the night sky in which certain constellations are visible all year round. The table given below indicates the minimum declination of circumpolar stars in various latitudes.

Lat.	Decl.	Lat.	Decl.
60° N	30° N	40° S	50° S
50° N	40° N	30° S	60° S
40° N	50° N	20° S	70° S
30° N	60° N	10° S	80° S

The map should be made as large as possible, preferably on heavy drawing paper or stiff card. Parallels of declination are drawn as a series of equidistantly spaced concentric circles: the scale is optional but if, for example, 2 inches (5 cm) is made to represent 10° then the radius of the 80° parallel would be 2 inches (5 cm); that of the 70° parallel, 4 inches (10 cm); and so on until the limiting circumpolar declination is reached. The central point is the celestial pole, from which a series of radial lines are drawn at 15° intervals. These are hour lines of Right Ascension, to be numbered successively from 0 to 24 hours in a clockwise direction on a northern map; anticlockwise on a southern map. Upon completion of the grid system, insert the brighter stars in their correct positions. Their co-ordinates of Right Ascension and declination can be obtained from Table 2.3, the Index Maps, or Whitaker's Almanack. The night sky should be observed as often as possible, fainter stars being added until a comprehensive map of the circumpolar constellations is built up.

4. AN EQUATORIAL STAR MAP

This map is designed to cover the region of the sky between the parallels of declination 50° N and 50° S, through the full range of R.A. from 0 to 24 hours. The network of meridians and parallels is similar to that used in the construction of the map described in Project 1.4, except for a difference in size. In this case the scale is larger: 1 inch (2 cm) to 1 hour of R.A., and $\frac{8}{10}$ inch (1·5 cm) to 10° of declination, giving an overall size of 24 inches (48 cm) by 8 inches (12 cm). As in the preceding project, the idea is to plot the positions of the brighter stars, fainter ones being inserted later, when practical observations have been made. Plotting the curve of the ecliptic on the map is a useful exercise too, because it demonstrates the eastwards motion of the sun with respect to the stars; it also helps one to become acquainted with the zodiacal constellations, the domain of the planets.

Maps made on black card, showing the stars depicted in white, or their characteristic colours, look particularly effective.

5. THE ORION REGION

The neighbourhood of Orion is so rich in bright stars as to make it worthy of special consideration. The region to be charted extends from R.A. 3 hours (right edge) to R.A. 8 hours (left edge), and from declination 40° N (upper edge) to declination 20° S (lower edge). Using a scale of 3 inches (6 cm) to 1 hour of R.A., and 2 inches (4 cm) to 10 degrees of declination, the map will measure 15 inches by 12 inches (30 cm by 24 cm). Lines of R.A. and declination are spaced 1 inch (2 cm) apart, that is every 20 minutes and 5 degrees respectively. Further subdivisions may be inserted along the edges of the map. On completion of the graticule, plot the positions of the following stars taken from Table 2.3: Betelgeuse, Bellatrix, Rigel, Saiph, Alnitak, Alnilam, Mintaka, Iota, Castor, Pollux, Procyon, Sirius and Aldebaran. When this has been done, go outside and compare your map with the sky around Orion. The stars which have been plotted will be recognised almost immediately and should be used to estimate the positions of fainter stars for insertion in the map. Did you notice the redness of Betelgeuse and Aldebaran, and the brilliant whiteness of Rigel? Are the V of the Hyades and the compact cluster of the Pleiades (R.A. 3 h 44 m, declination 24°·0 N) featured on your map? How many stars in the Pleiades can you see with the naked eye? Your map should show the Sword of Orion, comprising several fainter stars lying on a line through the central star of Orion's Belt and Iota Orionis. If a small telescope or binoculars is available, examine the Sword and see if you can find a hazy patch of light in the vicinity of Iota. This is the Great Nebula in Orion, a luminous gas cloud whose light takes over a thousand years to reach us (see Chapter 7). Some people can even detect the nebula with the naked eye.

6. THE PLOUGH

Observers to whom The Plough is circumpolar will find its changing aspect in the night sky a challenging problem for research; for its position, in common with all other stars, depends on the hour of night and the time of year. On a large disc of stiff card, draw a system of radial lines 30° apart and make a small hole at the central point. Sight the Pole Star through this hole, adjusting the disc until one of the radial lines coincides with the direction of the "pointers" of the Plough. Mark this line clearly. The edge of the disc which lies vertically over the pole must also be marked so that the card can be reset in the same position on successive observations. Two hours later The Plough will have swung round in an anticlockwise direction, the pointers now lying on the extension of the next radial line. Further observations will confirm that the pointers move over a radial line every two hours. Thus the diurnal motion of The Plough is seen to be performed at the rate of 30° every 2 hours, or 15° per hour, which is the rate of rotation of the celestial sphere due to the earth's rotation. This means that The Plough will traverse a quarter of its daily path, or 90°, in 6 hours. For example, on November 7 at 8 P.M. local time (9 P.M. British Standard Time), The Plough will be found below the Pole Star, with the pointers vertical; 6 hours later it will be to the east, standing on its handle with the pointers horizontal. Of course, The Plough will not be found in the same position at the same time every night, for the simple reason that the earth's revolution around the sun gives the stars an apparent annual motion in addition to their diurnal motion. This causes the stars to return to the same position in the sky 4 minutes earlier each night which, as we have seen, adds up to 2 hours each month. If the position of The Plough is noted on November 7 at 8 P.M. as described above, then it will be in the same position on December 7 at 6 P.M.; January 7 at 4 P.M. and, working backwards, October 7 at 10 P.M. September 7 at 12 P.M.; and so on. The lined disc will show the annual motion easily enough: observe The Plough at the same time of night at monthly intervals and it will be seen that as each month passes so the pointers move anticlockwise to the next radial line on the disc.

Keep records of your observations throughout the year, and get to know the position of The Plough at different times of night; the knowledge gained can be used to construct a star clock as described in Project 1, Chapter 4.

Incidentally, when you were observing The Plough, did you notice a faint star very close to Mizar, the second star from the end of the handle? This is Alcor. Mizar and Alcor together form what is known as a visual double, and people possessing average sight can usually see it as such.

7. THE MYTHOLOGICAL CONSTELLATIONS

By now you will be familiar with many of the constellations which adorn the night sky. Seek out a reference book on mythology in your local library and discover for yourself the fascinating legends behind the names of some of these constellations. In Greek mythology, for example, King Cepheus and Queen Cassiopeia had their beautiful daughter Andromeda chained to a rock in order to appease the wrath of the sea god who sent a monster to devour her. Perseus slew the monster, freed Andromeda and, taking her for his wife, carried her off to happiness on the winged horse Pegasus.

8. PHOTOGRAPHY

The constellations can be photographed quite easily by following the advice given in Project 5, Chapter 1. In this case, however, exposures are to be relatively short; between 3 and 10 seconds should be sufficient. Longer exposures giving extended star trials are also very interesting (see Plate I). Orion and The Plough are particularly good subjects to begin with.

ANSWERS TO REVIEW QUESTIONS

2. Mag. —1 is 6.3 times brighter than mag. +1
7. No.
8. Alioth. Vega. Deneb. Fomalhaut. Kaus Australis. Shaula.

CHAPTER 3

OUR HOME IN SPACE

The earth is one of a family of nine planets and numerous smaller bodies orbiting around the sun, known collectively as the solar system. Of the planets, the earth is third in order of distance from the sun and fifth in order of size.

THE EARTH

SHAPE AND SIZE

The earth is commonly regarded as a sphere, but strictly speaking it is slightly flattened at the poles and bulged at the equator; a shape more correctly defined as an *oblate spheroid* or an *ellipsoid*. Modern investigations involving the study of the movements of artificial satellites indicate that the earth is very slightly pear-shaped. Photographs taken from spacecraft clearly show the curved form of the earth: yet long before the dawn of the space age its curvature was evident to early astronomers from the shadow of the earth cast upon the moon during a lunar eclipse, when the earth comes between the sun and the moon. Further evidence stems from the fact that the earth can be circumnavigated, and anyone living near the sea may observe ships appearing and disappearing over the horizon. In navigation, the unit of distance is the *international nautical mile*, 6076·6 ft (1852 metres), which is the average length of one minute of arc of a meridian in latitude 45°. The unit on land is the *statute mile*, established in the reign of Queen Elizabeth I as 5280 ft. In astronomy, distances expressed in miles refer to statute miles. Metric measurements are now in common use and the basic unit of distance in the Metric System is the *metre*, originally defined as one ten-millionth of the distance along a meridian from the north pole to the equator. A kilometre (km) is one-thousand metres. The dimensions of the earth are:

equatorial diameter	7927 miles (12 757 km)
equatorial circumference	24 902 miles (40 075 km)
polar diameter	7900 miles (12 714 km)
polar circumference	24 860 miles (40 008 km).

THE EARTH'S INTERIOR

Beneath its surface, the earth is divided into three roughly concentric regions called the *crust*, the *mantle*, and the *core*.

The crust is the thin outermost shell, ranging in thickness from 3 to 5 miles (5 to 8 km) under the oceans to 12 to 56 miles (19 to 90 km) under the continents. To a large extent it is composed of igneous rocks such as granite and basalt, in parts overlain by sedimentary rocks such as sandstone and limestone. About 70 per cent of the earth's surface is covered with water, the remaining portion being made up with continents and islands. The highest point on land is regarded as the summit of Mount Everest at 29 028 ft (8488 m) above sea level. The lowest point of the ocean is in the Marianas Trench, off the Philippine Islands, at 36 198 ft (11 033 m) below sea level.

The mantle extends downwards to about 1800 miles (2897 km) below the crust and is mainly composed of heavy magnesium iron silicates: the upper part is rigid but the lower part is of a plastic consistency.

The core is the central region of earth below the mantle. The outer core, which is about 1300 miles (2092 km) thick, behaves as a liquid and is thought to consist of molten iron. The inner core, about 800 miles (1287 km) in thickness, is regarded as solid and is thought to consist mainly of iron and nickel.

From the study of radio active decay in crustal rocks, the age of the earth's solid surface is estimated at about 4500 million years.

THE EARTH'S ATMOSPHERE

The atmosphere is a mixture of gases surrounding the earth, extending from the surface to several hundreds of miles in height. In its lower regions it is composed of 78 per cent nitrogen, 21 per cent oxygen, nearly one per cent argon, and traces of carbon-dioxide, neon, helium, hydrogen, krypton and xenon. It also contains water vapour in variable amounts, together with large quantities of dust and salt particles. The atmosphere has weight and exerts a pressure of about $14\frac{1}{2}$ pounds per square inch (1 kg/cm^2) on the earth's surface. However, with increase of height pressure decreases because the air thins out very rapidly, nearly two-thirds of the mass of the atmosphere being contained below the height of Mount Everest. The thread supporting life on our planet is indeed a very slender one; a fact brought home to us very dramatically in the 1968 Olympic Games, when many athletes suffered from oxygen deficiency in the rarer atmosphere of Mexico City, which is only 8000 ft (2438 m) above sea level!

Although scientists have divided the atmosphere into a number of regions, as shown in fig. 3.1, we need only concern ourselves with a few of its more important features.

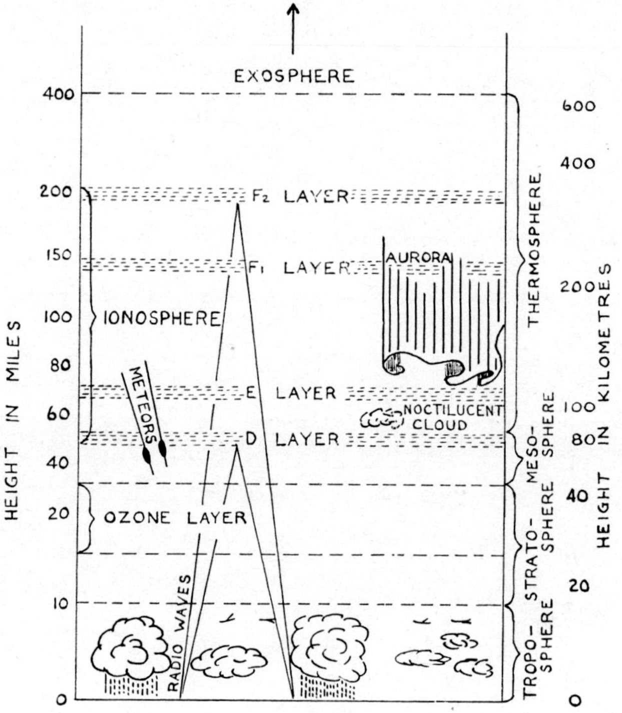

Fig. 3.1
Section through the Earth's Atmosphere.

The *troposphere* contains most of the water, either in the form of invisible water vapour or in the visible form such as clouds. The air is in a continuous state of turbulence and it is in this region that the atmospheric disturbances we call weather occur. Even a beautifully clear night can be spoiled for telescopic observation due to turbulent air currents; when the stars appear to be twinkling it is a sure sign that "seeing" is poor.

An important feature of the *stratosphere* is the existence, in its upper region, of a layer of ozone gas which acts like a massive umbrella, protecting us from harmful ultraviolet radiation coming our way from the sun.

In the upper atmosphere lie several layers having the property of being able to reflect radio signals from the earth back to the earth again, thus making world-wide radio communications possible. This "radio region", known as the *ionosphere*, is often disturbed by bursts of energy from the sun which disrupt radio communications and give rise to spectacular displays of *aurorae*. Sometimes, short-lived bright trails are seen across the night sky; these are *meteors* or, as some people call them, "shooting stars". They are small particles from outer space, burning up in the heat generated by friction with the earth's atmosphere. We shall be mentioning aurorae and meteors again later on.

The exosphere is the outermost region of the earth's atmosphere where the extremely rarefied air fades into interplanetary space.

THE EARTH'S MAGNETIC FIELD

The magnetic field of the earth is somewhat similar to that which would result from a powerful bar magnet placed at its centre and inclined at an angle of about 12° to the earth's axis of rotation. The north magnetic pole is situated in northern Canada, a little over 1000 miles (1600 km) from the north terrestrial pole, whilst the south magnetic pole is situated in Antarctica, about 1500 miles (2400 km) from the south terrestrial pole. Roughly mid-way between the magnetic poles lies the magnetic equator, an irregular line separating the northern and southern magnetic hemispheres, along which the direction of the field is horizontal. The field strength at the magnetic equator is about half that at the magnetic poles.

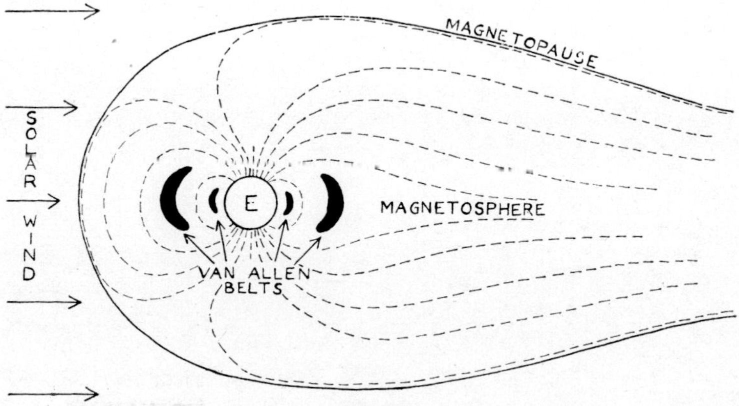

Fig. 3.2
The Earth's Magnetic Field—The Magnetosphere.

If a magnetised needle is freely suspended by a thread attached to its centre it will align itself in the magnetic meridian, *i.e.* the direction of the earth's field at the place. At places other than the magnetic equator the needle will be inclined to the horizontal by an angle called the magnetic dip or inclination. At the magnetic poles this inclination is 90°. The end of the needle directed towards the magnetic north is called the north seeking pole. At most places the direction of magnetic north and true north are different, the horizontal angle between the true and magnetic meridians being called the magnetic variation or magnetic declination. Variation is named east at places where the magnetic north lies to the right of the true north, and west where it lies to the left. Magnetic variation is subject to a number of changes, the most notable of which is called secular change. It is caused by movements within the core of the earth, resulting in a slow movement of the magnetic poles around the terrestrial poles. For example, at London the variation was 11° E in the year 1580; zero in 1657 and 24° W (maximum) in 1815. It is decreasing at the present time and is expected to be zero again in the early part of the 20th century. The property of a magnet to align itself in the magnetic meridian provides the basic requirement for a magnetic compass, though its usefulness deteriorates in high magnetic latitudes where the earth's field becomes more nearly vertical.

Beyond a few thousand miles from the earth the magnetic field becomes very weak and irregular. The sun is continuously pouring out energy in great gusts, called the *solar wind*, which exerts pressure on the magnetic field and distorts its outer regions beyond any resemblance to the field associated with a bar magnet. At distances of about 2000 and 10 000 miles (3200 and 16 000 km) from the earth lie two zones known as the *Van Allen belts*, where dangerous radiation from the sun has become trapped in the magnetic field. The whole volume of space occupied by the earth's magnetic field is called the *magnetosphere*.

GRAVITATION—THE EARTH'S GRAVITY

A stone swung round on the end of a length of string is kept in a circular path by the tension in the string. Should the string break, the tension in it becomes zero and the stone will fly off in a straight line in the direction it was moving at the instant the string snapped. Suppose we now consider the moon as it revolves around the earth, or the planets as they revolve around the sun. These bodies are kept in their assigned paths, or orbits, not by the tension in a string but by an invisible force called *gravitation*. The force of gravitation acts between two bodies, each body attracting the other. Sir Isaac Newton stated in his famous law "the force between any two bodies in the universe is directly proportional to the product of their masses and inversely proportional to the square of the distance between their centres".

From this statement the force between two masses is greater if the product of their masses is greater. That the force varies inversely as the square of the distance means it decreases rapidly as the distance is increased. For example, if we assume the force between the earth and the moon to be standard, then if the distance is doubled the force is reduced to $\frac{1}{4}$ of the standard force; if it is trebled the force is reduced to $\frac{1}{9}$, and so on.

A pendulum freely suspended in the vicinity of a large mountain is attracted out of the vertical by the much greater mass of the mountain by an extremely small but measurable amount. Similarly, because of its much greater mass, the earth attracts other smaller bodies towards its centre and it is said that Newton formulated his ideas about gravity after observing the fall of an apple from a tree. Whether or not this is true does not detract from the everyday experience that near the earth free objects fall downwards not upwards. The force of gravity causes a freely falling body to accelerate, *i.e.* increase its speed continuously while it is in motion. Near the surface the acceleration due to gravity, denoted by the letter *g*, is about 32·2 ft (9·81 m) per second^2. In order to appreciate this, imagine a stone dropped over the edge of a 150 ft (45·72 m) high vertical cliff: after one second has elapsed the stone's velocity is 32·2 ft/sec (9·81 m/s); after two seconds 64·4 ft/sec (19·62 m/s); and after three seconds 96·6 ft/sec (29·43 m/s).

Since the force of gravity depends on the distance from the earth's centre it follows that *g* must be greater at the poles than at the equator because of the smaller polar radius. Thus we can distinguish between mass and weight: The mass of a body is the amount of matter in it, whereas the weight of a body is governed by the attraction the earth has for it. Therefore, the mass of a given object is constant, but if it weighed say 200 lb (92 kg) at the equator, it would weigh about 201 lb (92·5 kg) if taken to the poles. Nowadays, one commonly hears of astronauts in space being referred to as weightless: they are still the same persons as they were before "lift off" but because the motion of the spacecraft is such as to nullify gravity they weigh nothing.

THE EARTH IN MOTION

The earth has several distinct motions in space, of which the two most important are a daily rotation on its axis and an annual revolution around the sun.

THE EARTH'S ROTATION

The earth rotates on its axis once per day, the motion being from east to west, or anticlockwise looking down on the north pole. Until the 16th century it was widely accepted that the earth was fixed and all other celestial bodies revolved around it. Nicholaus Copernicus is credited with first suggesting the earth's rotation, but it was not until the 19th century that it was conclusively proved by the French physicist Foucault by means of the now famous pendulum experiment, the first public demonstration of which took place in Paris in 1851. A large bronze sphere weighing over 50 lb (23 kg) was freely suspended from the dome of the Pantheon by a wire 200 ft (61 m) long. The pendulum was set in motion and it was arranged that a steel pin attached to the underside of the sphere just scraped the surface of a tray of sand. As time went by it became evident from the furrows cut in the sand that the pendulum plane was rotating in a clockwise direction. However, since no forces were acting to turn the pendulum plane, it was clear that the pendulum plane itself was fixed in space while the earth was turning beneath the pendulum in an anticlockwise direction.

The time for a complete rotation of the pendulum plane at the poles is 24 hours, *i.e.* at a rate of 15° per hour. In theory, the hourly rate in any latitude is 15° × sine latitude. Thus the time for a complete rotation of the pendulum plane in Paris is about 32 hours, and in London just over 30½ hours. At the equator the rate would be zero. In the southern hemisphere the pendulum would deviate in the opposite direction.

THE EARTH'S REVOLUTION

Copernicus is also credited with first suggesting the sun as the

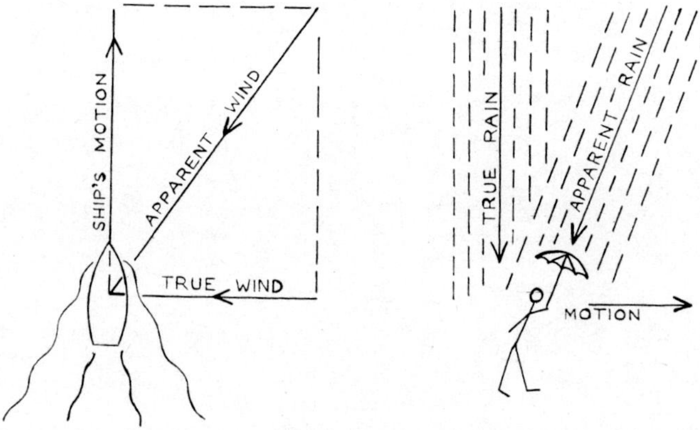

Fig. 3.3
Displacement of Wind and Rain in the Direction of the Observer's Motion.

central body of the solar system, with the planets revolving around it. Unfortunately, as in the case of the earth's rotation, he was unable to provide any proof of his theory. It may be thought that the eastwards motion of the sun with respect to the stars is proof of revolution, but this movement would in fact be evident even if the sun revolved around the earth. In spite of this, there is evidence for revolution and two of the main effects are: (1) *aberration of starlight*, (2) *parallax of the stars*, which is dealt with in Chapter 7.

Aberration, the apparent displacement of a star in the direction in which the earth is moving, was discovered by the English astronomer James Bradley in 1727. Similar effects may be observed on a moving ship, where the motion of the vessel causes the wind to appear to come from a direction nearer to the direction of motion; or where vertically falling rain appears to come from in front of a moving person (fig. 3.3). As the earth moves in its orbit, aberration is greatest when it is moving at right angles to the starlight and zero when moving parallel to it. At opposite points in the orbit, aberration is first in one direction and then in the other, so that during the course of a year each star appears to trace out a small orbit, the radius of which is 20·5 seconds of arc. Because the figure of 20″·5 was found to be constant for every star, the constant speed of light (186 000 miles or 300 000 kilometres per second) was verified.

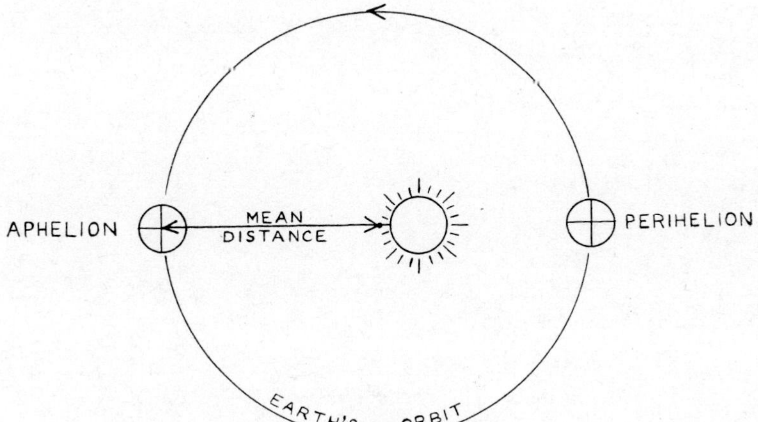

Fig. 3.4
The Earth's Orbit.

The earth's orbit around the sun is an ellipse of small eccentricity with the sun at one focus of the ellipse. The mean distance of the earth from the sun is 92 900 000 miles (149 600 000 km). This distance, known as the *astronomical unit*, is often used when stating the distances of the nearer astronomical bodies such as the planets. The earth is nearest to the sun in early January when it is said to be at *perihelion*. In July, when it is furthest away from the sun, it is said to be at *aphelion*. The perihelion and aphelion distances are respectively 1·5 million miles (2·4 million km) less and greater than the mean distance.

THE SEASONS

The earth revolves around the sun once in every year, giving rise, as we have seen, to the apparent annual motion of the sun around the ecliptic. The interval between two successive returns of the sun to the Vernal Equinox is called the *tropical year*, the year of the seasons, and is 365 days 05 hours 48 minutes and 46 seconds in length.

CAUSE OF THE SEASONS

The earth's axis is inclined $23\frac{1}{2}°$ to the perpendicular to the plane of its orbit. As the earth revolves around the sun it keeps its

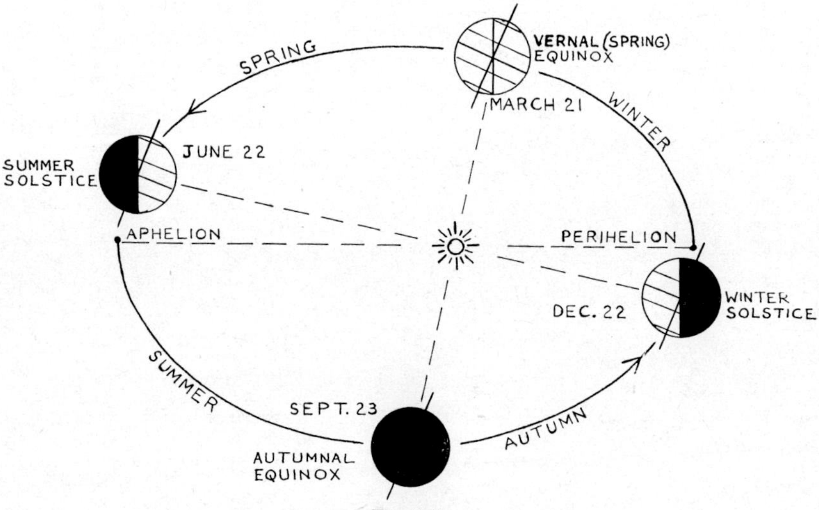

Fig. 3.5
The Seasons.

axis directed to the same point in space. Thus, each pole is presented towards the sun for part of the year and turned away from it for the remainder of the year. To get the true picture hold out the clenched left hand to represent the sun, and take a pencil in the right hand, inclined out of the vertical to represent the earth's axis. Now, with the left hand fixed, revolve the right hand anti-clockwise around it without altering the inclination of the pencil, keeping its upper end pointed to the same general direction.

The combination of the earth's axial tilt and its orbital motion gives rise to the four seasons—Spring, Summer, Autumn and Winter. The relation between the seasons and the sun's ecliptical position is given in the table below.

Northern Hemisphere	Southern Hemisphere	Approximate Beginning	Duration	Sun's Ecliptic Position
Spring	Autumn	March 21	92d 20h	R.A. 0h, Decl. 0°
Summer	Winter	June 22	93d 14h	R.A. 6h, Decl. $+23\frac{1}{2}$°
Autumn	Spring	Sept. 23	89d 19h	R.A. 12h, Decl. 0°
Winter	Summer	Dec. 22	89d 01h	R.A. 18h, Decl. $-23\frac{1}{2}$°

The seasons are of unequal length because the earth's orbital speed is changing continuously, being faster at perihelion and slower at aphelion (see laws of planetary motion, Chapter 6).

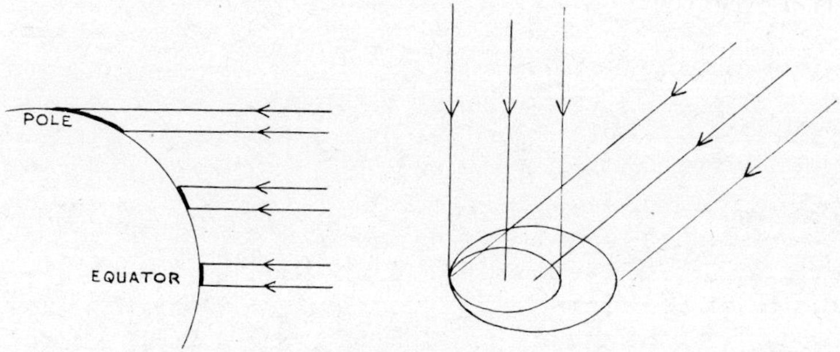

Fig. 3.6
Variation in the Sun's Heating Effect.

The change in the sun's declination during the course of the year causes a day to day variation in its noon-day altitude, leading in turn to a variation in the amount of heat received from the sun. It is convenient to think of the energy radiated from the sun as travelling in straight lines called rays, and the sun distant enough for these rays to reach the earth more or less parallel to each other. If we consider a bundle of rays falling on a part of the earth, the heating effect is distributed over a smaller area when the altitude is large than when the altitude is small (fig. 3.6). This explains why it is warmer in summer than winter and why the equatorial regions are warmer than the polar regions.

The region between latitudes $23\frac{1}{2}°$ N (Tropic of Cancer) and $23\frac{1}{2}°$ S (Tropic of Capricorn) experiences the midday sun at the zenith on two occasions during the year, once when the sun's declination is increasing and again when it is decreasing. This region, called the *torrid zone*, is characterised by high temperatures and a small annual range of temperature.

The regions extending $23\frac{1}{2}°$ from the poles to lat. $66\frac{1}{2}°$ N (Arctic Circle) and lat. $66\frac{1}{2}°$ S (Antarctic Circle) are called the *frigid zones*. They are characterised by low temperatures and a small annual range. At any place within these zones the sun will become circumpolar; hence the expression "land of the midnight sun". When a particular latitude is experiencing continuous daylight the same latitude in the opposite hemisphere will be experiencing continuous darkness. It may now be appreciated that since the sun spends six months in each celestial hemisphere, the north and south terrestrial poles are alternately in daylight for one half of the year and in darkness for the other half, although the darkness will not be total for the full six months due to the effects of twilight, mentioned later.

The regions in each hemisphere between the torrid and frigid zones, in which the sun neither becomes circumpolar nor directly overhead, are called the *temperate zones*. Within them, seasonal variations of temperature are of great variety, some places experiencing hot summers and extremely cold winters while other places have cool summers and mild winters.

The large land masses heat and cool very quickly with change of season, leading to more extreme seasonal effects, whereas places near the large oceans have their climates modified by the presence of water, which does not gain and lose heat so readily as the land. Large modifications occur in some instances where ocean currents bring warmer water from the tropics into the temperate latitudes, a good example being the tempering effect of the warm Gulf Stream on the climate of the British Isles. Compare the British climate with the extreme climate of places in Canada having the same latitude.

There is a tendency for the seasons to lag behind the sun due to the blanketing effect of the earth's atmosphere. There is also a tendency for the seasons to be less extreme in the southern hemisphere due to the modifying influence of the large extent of the oceans in that hemisphere, sufficient to outweigh the effects of the earth being closer to the sun during southern summer and further away during winter.

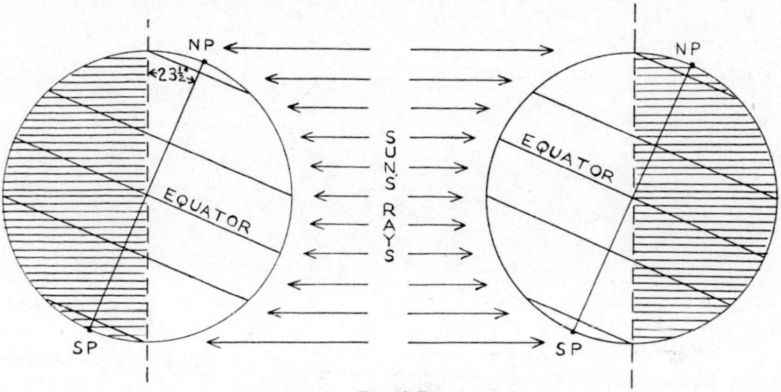

Fig. 3.7
Unequal Length of Day and Night.

The change in the sun's declination also brings a seasonal variation in the lengths of day and night. The duration of daylight exceeds that of darkness in the northern hemisphere when the sun is in north declination, the longest days and shortest nights occurring at the summer solstice. The duration of darkness exceeds that of daylight when the sun is in south declination, the longest nights and shortest days occurring at the winter solstice (fig. 3.7). The reverse is true in the southern hemisphere. The lengths of day and night are always equal at the equator and equal all over the earth at the equinoxes.

Twilight is the period of "part-light" occurring before sunrise and after sunset, caused by the scattering of the sun's light by dust and moisture in the earth's atmosphere. Morning twilight (dawn) lasts from the time the sun is 18° below the horizon until sunrise, and evening twilight (dusk) from sunset until the sun sinks 18° below the horizon. At the equator, twilight is short because the sun's diurnal path cuts the horizon at a steep angle and consequently the sun passes through the twilight zone very quickly. In progressively higher latitudes twilight becomes longer because the sun crosses the horizon at a smaller angle and takes longer to pass through the twilight zone. The changing declination of the sun also affects the duration of twilight, which in temperate and high latitudes is noticeably longer during the summer months.

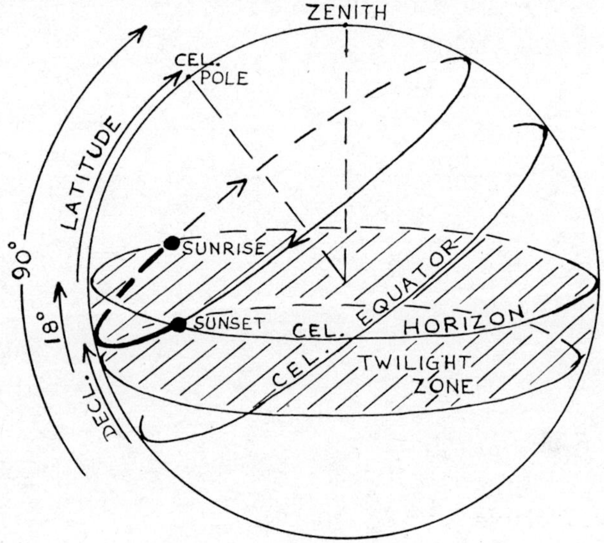

Fig. 3.8
Conditions for Continuous Twilight.

A celestial body becomes circumpolar at any place when the latitude and declination have the same name, and their sum is more than 90°. This means the sun, whose maximum declination is 23½°, can only become circumpolar in places having a latitude greater than 66½° N and S. Twilight is uninterrupted by an interval of complete darkness when the latitude and declination have the same name and their sum plus 18° is more than 90°, or when the declination exceeds 72° minus the latitude. This general rule is illustrated in fig. 3.8. For example, London (lat. 51½° N) experiences continuous twilight between sunset and sunrise when the sun's declination is greater than 20½° N. Reference to Table 1.1 shows this occurs between about May 23 and July 21.

THE EARTH IS OUR OBSERVATORY

From our small planet, spinning and revolving in space, our knowledge of other celestial bodies is derived mainly from visible light which travels through space with a speed of 186 000 miles per second (300 000 km/s).

LIGHT AND THE TELESCOPE

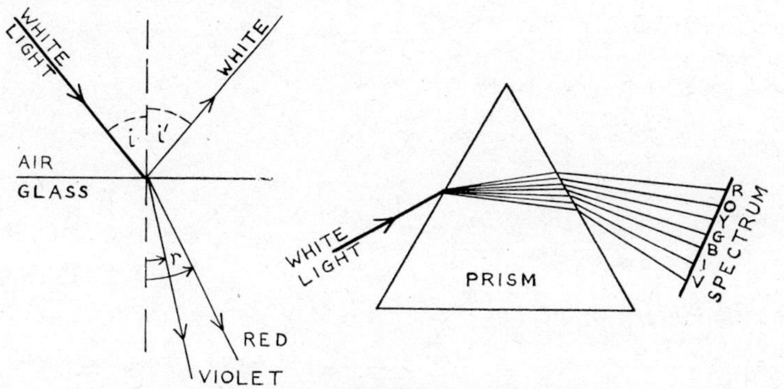

Fig. 3.9
Reflection, Refraction and Dispersion of Light.

If a narrow beam of ordinary white light falls obliquely upon a flat glass block, the angle between the beam and the perpendicular, or normal, to the plane of the glass is called the angle of incidence. Firstly, part of the incident light is reflected from the glass such that the angle of incidence equals the angle of reflection. Secondly, when the beam enters the glass it is bent or refracted towards the normal. Refraction is always towards the normal when the light passes into a medium of greater density, and away from the normal when the medium is of lesser density. Thirdly, dispersion takes place because white light is a mixture of seven colours—violet, indigo, blue, green, yellow, orange and red—and each is refracted by a different amount, being greater for the violet and least for the red. Therefore, refraction is always accompanied by dispersion.

Telescopes, used to facilitate a more detailed examination of distant objects than is possible with the naked eye, are of two main types: (1) *refractors* which employ lenses; (2) *reflectors* which use mirrors.

A lens is a piece of glass ground and polished with one or both surfaces curved, its function being to intercept light rays from an object, refract them and bring them to a focus. The distance from the centre of the lens to the focus is its *focal length.* In 1608 a Dutch spectacle maker, Hans Lippershey, discovered the principle of the refracting telescope by using two lenses in combination. When Galileo heard about the Dutchman's invention he was quick to realise its potential in the field of astronomy. With small telescopes of his own making, some of which are preserved in museums at Florence, Galileo opened up one of the most exciting chapters in the history of astronomical discovery. A simple

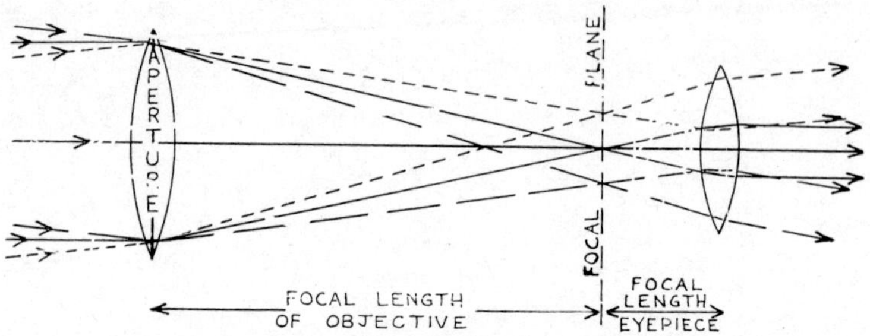

Fig. 3.10
Principle of the Refracting Telescope.

astronomical refractor consists of two double-concave lenses separated by a distance equal to the sum of their focal lengths. The lens which first receives the light from a distant source is the *objective* and the other, through which the image is observed, is the *eyepiece*. Of the two lenses the objective has the longer focal length and its diameter is called the *aperture* of the telescope. The eyepiece is set in a sliding tube so that the observer can adjust it until a sharp image is obtained. Unfortunately, refractors suffer from *chromatic aberration* caused by dispersion of light as it passes through the lenses; the different colours do not come to the same focus, resulting in blurred images. Chromatic aberration can be reduced by increasing the focal length of the objective, and in the early days this led to the construction of mammoth telescopes which were so cumbersome that they were not only difficult to control but often tended to sag under their own weight. In the 18th century the invention of the *achromatic lens*, a two-part lens of different curvatures and different types of glass, solved the

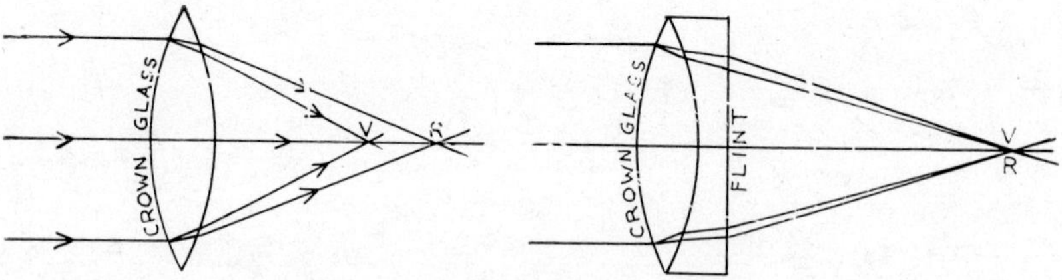

Fig. 3.11
Cause and Correction of Chromatic Aberration.

problem. Nevertheless, the difficulty of supporting large objectives, coupled with the problem of making large lenses free from defects, contributed to the limitation in the size of refractors. The largest one in the world is the 40-in (1·016 m) at Yerkes Observatory, Wisconsin, USA.

Since curved mirrors are free from the effects of chromatic aberration, the majority of astronomical telescopes constructed nowadays employ mirrors rather than lenses. However, in order to bring the rays of light from all parts of a mirror to the same focus, its curvature must be of a special form called *paraboloid.* The first telescope using this principle was made by Sir Isaac Newton in 1668, but it was too small to be of practical use. One of the most notable contributors to the development of the reflecting telescope was Sir William Herschel, who in the 18th century built many large instruments, the finest of which was of 4 ft (1·22 m) aperture and focal length 40 ft (12·2 m). The mirrors were made of *speculum metal*, an alloy of two parts copper and one part tin, but as they tended to tarnish very easily they required frequent refiguring. Mirrors are now made of glass, coated with a thin film of aluminium which provides a highly efficient and durable reflecting surface; the glass simply provides the required shape for the aluminium coating. In the *Newtonian* type of reflector the converging light rays from the *primary mirror* are intercepted

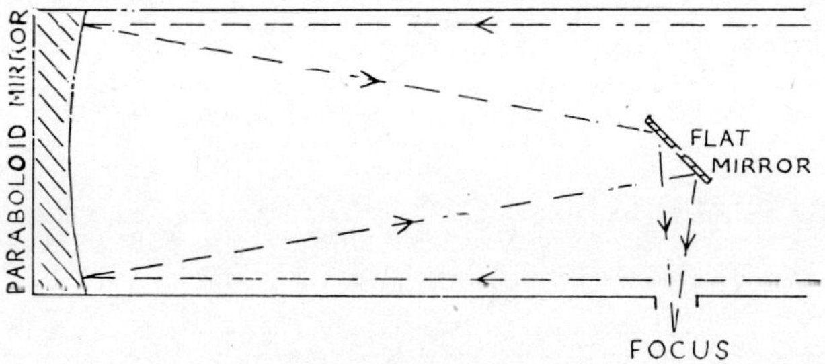

Fig. 3.12
Principle of the Newtonian Reflector.

by a small *flat mirror*, angled at 45° to the telescope axis, and reflected to the side of the telescope where they are brought to a focus for examination with the eyepiece. The image provided by an astronomical telescope is both inverted and reversed left to right. Photographs in books are often orientated to match the view through such a telescope, a fact which should be borne in mind when using binoculars or a terrestrial telescope. Although small refractors are in some respects superior to reflectors of similar aperture—the lens surfaces do not require periodic realuminising and they tend to give clearer images away from the centre of the field of view—there are other distinct advantages of reflectors over refractors: (1) minor internal defects which can render a glass disc useless for a lens may be tolerated in a mirror, provided the surface to be aluminised can be ground to the required shape without any of the defects breaking through; (2) a mirror has only one surface to be ground; (3) whilst a lens can only be supported around its circumference, a mirror can be mounted in a cell to give both edge and back support; (4) reflectors are generally of much shorter focal length than refractors of the same aperture so they are not so cumbersome; (5) mirrors are perfectly achromatic.

Telescopes of less than several inches apertures are usually mounted on a tripod so that they have freedom to be moved in the vertical plane (altitude) and in the horizontal plane (azimuth). Larger telescopes are more suited to the *equatorial mounting* in which the axis of the mounting is elevated in the meridian to the latitude of the place, so that it is directed to the celestial pole and aligned with the earth's axis of rotation. This allows the telescope to be turned parallel to the celestial equator in an east-west direction along the diurnal circle of a star. It is also free to move at right angles to the equatorial plane in a north-south direction, from one declination to another. A mounting of this type fitted with Right Ascension and declination scales permits an object to be located quickly from the co-ordinates given in a star atlas. An added refinement is an electric clock-drive which moves the telescope to the west in the equatorial plane at a speed sufficient to counteract the earth's rotation to the east, thus making it possible to hold a star in the field of view.

Numbered among the world's largest reflectors are: a 235-in (5·97 m) and a 102-in (3·05 m) in the USSR; the 200-in (5·08 m) Hale telescope on Mount Palomar, the 120-in (3·05 m) at Lick Observatory, and the 100-in (2·54 m) on Mount Wilson, all in California, USA; the 158-in (4·01 m) and the 90-in (2·28 m) Kitt Peak telescopes in Arizona, USA; and the 98-in (2·49 m) Isaac Newton telescope at the Royal Observatory, Herstmonceux, England. Other telescopes having apertures of 72 in (1·83 m) or more include those at the McDonald Observatory, Texas, USA; St. Michelle, France; Tokyo, Japan; the Dunlap Observatory, Ontario, Canada; Mount Stromlo, Canberra, Australia; the Ratcliffe Observatory, South Africa; and the Dominion Observatory, Victoria, B.C.

PROPERTIES OF OPTICAL TELESCOPES

The *magnifying power* of a telescope is the number of times the telescope increases the apparent diameter of an object compared with the naked-eye appearance.

$$\text{magnification} = \frac{\text{focal length of objective}}{\text{focal length of eyepiece}}$$

Example 1.

A telescope with an objective lens of focal length of 48 in (1219 mm) is supplied with two eyepieces of focal lengths $\frac{1}{2}$ in (12·7 mm) and $\frac{3}{4}$ in (19·05 mm) respectively. What magnification will each eyepiece give?

For the $\frac{1}{2}$-in eyepiece: magnification $= 48 \div \frac{1}{2} = 96$.
For the $\frac{3}{4}$-in eyepiece: magnification $= 48 \div \frac{3}{4} = 64$.

In this and other examples the reader is invited to check the results using the metric equivalents.

It has been found that there is little to be gained by using a magnification in excess of 50 times the aperture in inches, or twice that in millimetres, because the undue spreading of the light gives a weaker image. Therefore, if the aperture of the telescope in the above example is 4 in (101·6 mm), the maximum useful magnification is about 200.

Stars are so far away that they always appear as points of light no matter what magnification is used. For them, *light-gathering power* is far more important than magnification, the brightness of the image being directly proportional to the area of the objective or the square of its diameter.

$$\text{Image brightness} \propto A^2.$$

This means that a larger diameter objective will render fainter stars visible than one of smaller aperture.

Example 2.

If the aperture of the lens of the human eye when fully dilated at night is taken as $\frac{1}{3}$ in (8·47 mm), how much brighter would a star just visible with the naked eye appear when observed through a telescope of 6 in (152·4 mm) aperture?

Image brightness of eye $\propto (\frac{1}{3})^2$.
Image brightness of telescope $\propto 6^2$.
Ratio of image brightness $= 6^2 \div (\frac{1}{3})^2$
$= 6^2 \times 9$
$= 324.$

Thus the telescope gives a star image 324 times brighter than the eye alone.

TABLE 3.1
IMAGE BRIGHTNESS OF STARS

APERTURE		IMAGE BRIGHTNESS COMPARED WITH NAKED EYE	THEORETICAL VISIBLE MAGNITUDE
inches	mm		
Naked eye		1	6
1	25·4	9	8
2	50·8	36	10
4	101·6	134	11
6	152·4	324	12
10	254·0	900	13
20	508·0	3600	15
50	1270·0	22 400	17
100	2540·0	90 000	18
200	5080·0	360 000	20

In practice, a loss of one or two magnitudes will probably occur due to light loss in the telescope and atmospheric absorption.

Extended objects such as the sun, moon and planets give a disc image in the focal plane of the objective, the area of which is dependent upon the square of its focal length.

$$\text{Image area} \propto F^2.$$

However, image brightness increases with aperture squared and decreases with image area, thus:

$$\text{Image brightness} \propto \frac{A^2}{F^2}.$$

Example 3.

Two telescopes have apertures of 4 in (101·6 mm) and 2 in (50·8 mm) and focal lengths 60 in (1524 mm) and 24 in (609·6 mm) respectively. They are used separately without eyepieces to project images of the sun on to white cards placed in the focal planes of the objective lenses. What difference will there be in the size and brightness of the images?

Image area of 4-in $\propto 60^2$.
Image area of 2-in $\propto 24^2$.
Ratio of image areas $= 60^2 \div 24^2 = 6\frac{1}{4}$.

Therefore, the image provided by the 4-in is $6\frac{1}{4}$ times as large as that given by the 2-in.

$$\text{Image brightness of 4-in} \propto \frac{4^2}{60^2}.$$

$$\text{Image brightness of 2-in} \propto \frac{2^2}{24^2}.$$

$$\begin{aligned}\text{Ratio of image brightness} &= \frac{4^2}{60^2} \div \frac{2^2}{24^2}\\ &= \frac{4^2}{60^2} \times \frac{24^2}{2^2}\\ &= \left(\frac{4}{5}\right)^2\\ &= \frac{16}{25}.\end{aligned}$$

Thus, the image brightness with the 4-in is less than that with the 2-in.

NOTE: If the telescope were used in the normal way to observe stars, the ratio of brightness would be $4^2 \div 2^2$, so that stars would appear four times brighter in the 4-in than they would in the 2-in.

The linear size of the image formed by an objective is given by:

$$\text{Image diameter} = \text{Angular diameter of object} \times \frac{\text{focal length}}{57{\cdot}3}.$$

Example 4.

Find the diameter of the image of the moon, whose angular diameter is $\frac{1}{2}°$, in the focal plane of an objective of focal length 60 in (1524 mm) and aperture 4 in (101·6 mm).

$$\begin{aligned}\text{Image diameter} &= \tfrac{1}{2} \times \frac{60}{57{\cdot}3}\\ &= 0{\cdot}52 \text{ in } (13{\cdot}2 \text{ mm}).\end{aligned}$$

The ratio of focal length to aperture is called the focal ratio or *f-number*, which is important in astronomical photography. Thus the telescope in the above example has a focal ratio of f15. Generally speaking, the higher the f-number the longer the exposure required in order to obtain good results. However, anyone keen to take up astrophotography seriously is strongly recommended to read a book specially devoted to the subject (see list of further reading).

Many stars which appear single to the naked eye are seen to be two or more stars when seen through a telescope. The ability to separate celestial objects, called *resolving power*, depends directly upon the aperture (A) of the telescope.

$$\text{Resolving power in seconds} = \frac{4{\cdot}56}{\text{A-inches}} \text{ or } \frac{115{\cdot}82}{\text{A-millimetres.}}$$

High resolving power is also necessary in order to clearly distinguish features of the moon which are close together; the same applies to the sun and the planets.

ATMOSPHERIC REFRACTION

On passing through the gradually thickening layers of the earth's atmosphere, the light from a distant celestial object is refracted so that by the time it enters the observer's eye it appears to come from a direction above the true direction of the object.

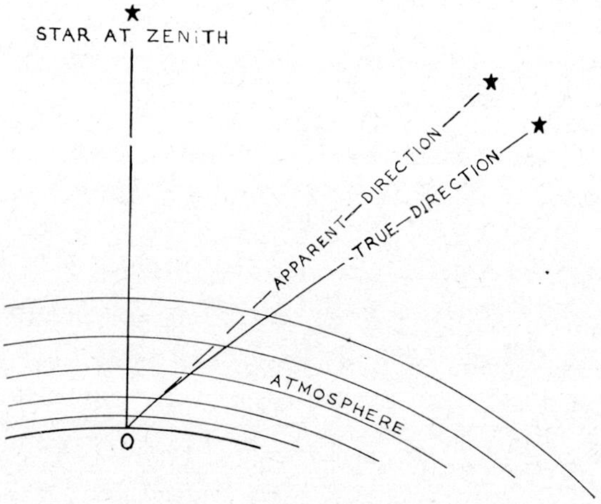

Fig. 3.13
Atmospheric Refraction Apparently Elevates a Celestial Object.

Therefore, atmospheric refraction causes any celestial body to appear at a greater altitude than it really is. Its value depends upon the angle the incoming light makes with the horizon: it is greatest when the body is at a low altitude when the light has to pass through a greater thickness of atmosphere, becoming less with increase in altitude. When a body is at the zenith, the light passes perpendicularly through the atmospheric layers and refraction is zero. The sun is about 32′ in diameter: when its lower limb is theoretically on the horizon it is apparently raised by 33′, while the upper limb is elevated by about 28′, resulting in its distinctly flattened appearance at sunset and sunrise. Refraction advances sunrise and retards sunset, giving a slightly longer duration of daylight than there should be theoretically. It also slightly extends the area where the midnight sun is experienced and reduces the area where it never rises.

The twinkling, or *scintillation*, of the stars is caused by variable refraction arising from turbulent air currents in the lower atmosphere. Planets do not always twinkle because, unlike the stars which are points of light, they have appreciable discs; although each point of the disc may be twinkling, the effects are out of phase and tend to cancel out.

REVIEW QUESTIONS

1. State one important feature of the troposphere, the stratosphere and the ionosphere.

2. At what points on the earth's surface will a freely suspended compass needle settle horizontally and vertically.

3. Give one or two words to match each of the following definitions:

   (a) the earth's daily spinning on its axis.

   (b) the annual motion of the earth around the sun.

   (c) the point in its orbit where the earth is furthest from the sun.

   (d) the mean distance of the earth from the sun.

4. The earth is farthest from the sun when it is summer in the northern hemisphere. Why, then, is it summer?

5. What would be the effect on the seasons and on variations in the length of day and night if the earth's axis was inclined 47° from the perpendicular to the plane of its orbit?

6. What is the essential difference between refracting and reflecting telescopes?

7. On a night when the stars were seen to be twinkling would the use of an eyepiece of greater magnification be likely to improve the quality of the images?

8. Which of the following telescopes would you prefer, assuming that they are both optically good and of equal aperture:

   (a) a refractor having an alt-azimuth mount on a tripod.

   (b) a reflector having an equatorial mount on a tripod?

PROJECTS AND RESEARCH

1. DETECTING THE EARTH'S ROTATION

The Foucault pendulum experiment provided the first convincing proof of the earth's rotation and similar experiments are often staged publicly at science museums, large planetaria and observatories. Although it is best performed along the original lines, the experiment can be carried out on a smaller scale, giving reasonable results. At a recent exhibition at the South Shields Marine and Technical College the local Astronomical Society used a pendulum consisting of a 40 ft (12 m) length of piano wire and a 20 lb (9 kg) iron sphere. The tray of sand was dispensed with and instead a system of radial lines one degree apart was marked out on the floor. A pointer, just clear of the floor, attached to the underside of the sphere coincided with the central intersection of the radial lines when the pendulum was at rest. The pendulum oscillated for about $\frac{1}{2}$ hour, during which time the pendulum plane apparently deflected about $5\frac{1}{2}°$, giving an hourly rate of 11°. This compared favourably to a calculated rate of 12°·3 per hour. In spite of this slight discrepancy, the rotation of the earth was clearly demonstrated.

When setting up the pendulum experiment the means of suspension should incur as little friction as possible. To set the pendulum swinging, draw the bob to one side and secure it with a light cord. When it is perfectly steady, burn through the cord.

2. EXPERIMENT TO DETERMINE THE ACCELERATION DUE TO GRAVITY (g)

The average value of *g* at the earth's surface is 32·2 ft/sec^2 or 9·81 m/s^2. An experiment making use of a simple pendulum, capable of giving results comparable to these figures is now described.

Apparatus—pendulum bob, cotton, pliers, stop watch, metre rule, retort stand and clamp.

Procedure—tie a length of cotton to the bob and suspend the cotton from the jaws of the pliers held in the clamp such that the length of the pendulum, measured from the point of suspension to the centre of the bob is exactly 1 m (100 cm). Place a piece of paper with a vertical mark on it behind the pendulum so that when the pendulum is at rest it hides the mark from an observer standing directly in front of it. Swing the pendulum through an arc, not exceeding 5°, starting the watch when the pendulum passes the vertical mark and stopping it 50 complete oscillations later when it passes the mark in the same direction. Repeat the procedure and record both times. Shorten the length of the pendulum successively by 5 cm down to 40 cm and take the times of 50 oscillations. Record the results in the form of a table as shown below.

Length of pendulum l in cm	Times for 50 complete oscillations in seconds 1st time	2nd time	Mean time t secs	t^2

Since 50 oscillations have been timed, then:

$$\frac{t}{50} = 2\pi\sqrt{\frac{l}{g}}$$

$$\text{and } t^2 = 2500 \times 4 \times \frac{22}{7} \times \frac{22}{7} \times \frac{l}{g}$$

$$\text{whence } g = 987\,75{\cdot}5 \times \frac{l}{t^2}\ \text{cm/s}^2 \text{ or } 987{\cdot}75 \times \frac{l}{t^2}\ \text{m/s}^2.$$

From this equation, values of g can be calculated for successive lengths and times of the pendulum, the mean value of all the results being taken as the acceleration due to the earth's gravity.

3. THE SEASONS

Prepare a dossier entitled "The Seasons—Their Causes and Effects", entering as much information as possible under several of the section headings given below.

(a) The Causes of the Changing Seasons.

(b) Day and Night—trace the change in the length of day and night from the times of sunrise and sunset given in newspapers or an almanac. At the equinoxes and solstices estimate the position of the rising and setting sun, either in terms of azimuth or with reference to fixed points on the landscape.

(c) Altitude of the Sun—estimate the altitude of the noon day sun, or, obtain it more accurately by means of the device described in Project 2, Chapter 1, or by shadow stick measurements.

(d) Weather—make a daily study of air temperatures (in the

shade), duration of sunshine, frequency and intensity of fog, rain, snow, gales and weather damage. Include newspaper and magazine cuttings.

(e) Landscape—changes in trees, shrubs and flowers. Obtain colour photographs of a particular scene during each of the four seasons.

(f) Animal, Bird and Insect Life—if possible select one of each kind for careful study.

(g) Personal Activity—the ways in which you adapt your way of life to the changing seasons, *e.g.* leisure activities, clothing, diet, etc.

(h) The Night Sky—list the major constellations visible at the same time of night during Spring, Summer, Autumn and Winter.

4. TWILIGHT

Record the times of sunrise and sunset at fortnightly intervals and estimate the duration of morning and evening twilight. Using the methods outlined earlier, determine whether some nights during the year are not completely dark in your locality and, if so, between what dates. Carefully record the colours of the sky around twilight, noting particularly vivid or pastel shades and taking occasional colour photographs. When sufficient data has been gathered, either confirm or deny the truth of the old adage "red sky at night—shepherd's delight, red sky in the morning—sailor's warning".

Investigate the list of places and latitudes given below, stating

(a) those where there is a period of complete darkness each night throughout the year.

(b) the dates between which some of the places do not experience complete darkness.

(c) the dates between which some of the places experience the midnight sun.

Plymouth, England	50°·4 N	Murmansk, Russia	69°·0 N
London, England	51°·5 N	Stockholm, Sweden	59°·3 N
Manchester, England	53°·5 N	Churchill, Canada	58°·8 N
Newcastle, England	55°·0 N	Anchorage, Alaska	61°·2 N
Edinburgh, Scotland	56°·0 N	Seattle, America	47°·6 N
Thurso, Scotland	58°·6 N	Capetown, S. Africa	33°·9 S
Aberystwyth, Wales	52°·4 N	Melbourne, Australia	37°·8 S
Belfast, N. Ireland	54°·6 N	Stanley, Falkland Is.	51°·8 S
Hammerfest, Norway	70°·7 N	McMurdo, Antarctica	77°·8 S

5. DEMONSTRATING DAY AND NIGHT

Using the knowledge gained in this chapter, construct models which may be suitably illuminated in a darkened room to demonstrate the varying length of day and night with the changing seasons.

6. FINDING MAGNETIC VARIATION

First of all it is necessary to determine the true meridian. This may be done by following the method outlined in Project 1, Chapter 1.

Pin a sheet of drawing paper to a board and draw in the true meridian. Then suspend a magnet by a thread attached to its centre so that the point of suspension is directly above the meridian line. Make sure there is no magnetic material in the vicinity, removing any keys etc. which may be lurking in pockets and any other metallic objects about your person to a safe distance. Now wait until the magnet is perfectly steady. Place a wooden ruler on the board and, looking down from directly above, line it up with the magnet. Draw in the magnetic meridian and measure the angle it makes with the true meridian. Repeat the measurement several times after deflecting the magnet and allowing it to settle each time. The mean of the observations can be accepted as the magnetic variation.

7. MAKING AN ELLIPSE

Every astronomical body moves in an orbit of some kind, the most common form of which is the ellipse. Lay a sheet of paper on a board and insert two pins a short distance apart. Place a knotted loop of string of length slightly more than this distance over the pins, drawing the loop tight with the tip of a pencil as shown in fig. 3.14. Move the pencil all the way round the pins, keeping the string tight at all times—the resulting figure is an ellipse and the positions of the pins are its foci. The longest diameter of the ellipse, passing through the pins, is the major axis; the shortest diameter, at right angles to the major axis through the centre of the ellipse, is the minor axis. An important property of the ellipse is that from any point on it the sum of the distances to the foci is constant. Eccentricity is the term used to express the amount of flattening of an ellipse and is defined as half the distance between the foci divided by half the major axis.

For the same length of string, ellipses of varying eccentricity may be drawn by altering the distance between the pins. Eccen-

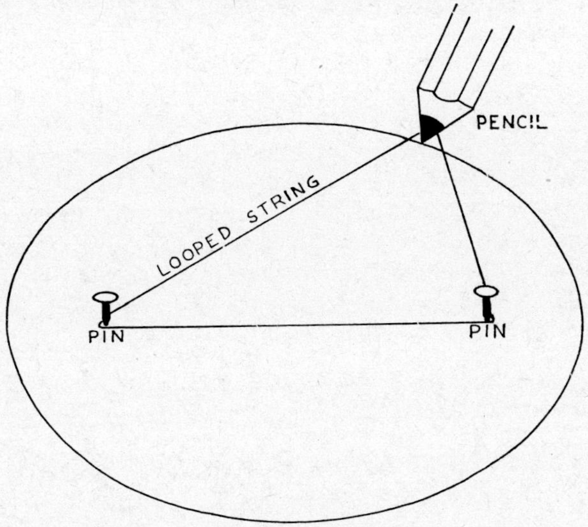

Fig. 3.14
Construction of an Ellipse.

tricity decreases as the distance between the pins is decreased, until, when the distance is zero, the eccentricity is zero and the resulting figure is a circle.

The earth's orbit around the sun is an ellipse of small eccentricity with the sun at one focus of the ellipse. In fact the eccentricity of the earth's orbit is only about 0·017, which means that it is not greatly different from a circle. To illustrate this, draw an ellipse having a major axis of length 186 units and a distance between its foci of three units.

8. ASTRONOMICAL OBSERVATORIES

An interesting project would be to gather as much information as possible about the world's largest observatories, including details of the design and size of telescopes, accompanied by drawings and photographs copied from scientific magazines and library books. Some observatories have prepared literature which is available on request. If you live near an astronomical observatory or an optical works you should enquire about the possibility of making a visit. This project may be combined with Project 9,

Chapter 7 and could include an account of the development of the optical and radio telescopes.

Actually it is not beyond the capability of most people to construct a telescope, given a fair amount of patience and determination. However, detailed guidance is necessary and the interested reader is advised to refer to a book specially devoted to the subject of telescope making (see list of further reading). Several manufacturers advertising in scientific magazines can supply telescope kits for assembly at home.

CHAPTER 4

TIME AND THE CALENDAR

Time is a very important aspect of our daily lives and, whether we like it or not, most of our activities are governed by the clock. We have become so used to this arrangement that we seldom think about what time really is. The dictionary defines time as "the interval between two successive events" and on the face of it this definition seems to be quite satisfactory. The trouble is that clocks can run fast or slow, so it becomes necessary to check them against time signals related to standard events, associated as closely as possible with our human activities. We have previously encountered two such standard events, namely, the earth's rotation on its axis and its revolution around the sun. These earthly motions provide us with our day and year respectively.

It is true that astronomers have detected slight irregularities in the rotational period, amounting to a few thousandths of a second, but for most purposes these variations are of little importance and time based on the earth's rotation is in everyday use. If a certain celestial object is selected as the "time reckoner" it becomes possible to measure the period of the earth's rotation by the interval between two successive transits of the "time reckoner" across a given meridian.

SIDEREAL TIME

The interval between two successive transits of any star across the same meridian is called a *sidereal day*. Strictly speaking, the sidereal day is determined by the apparent daily motion of the point on the celestial sphere known as the Vernal Equinox or First Point of Aries. At any place the sidereal day begins when Aries is on the superior meridian and is reckoned through 24 sidereal hours to the next upper transit.

A star is related to the First Point of Aries through its Right Ascension, making it possible for astronomers, using special instruments, to obtain the exact instant of a star's meridian transit in order to keep a check on sidereal time. We shall see later how time by the stars is related to our *Civil Time*.

APPARENT SOLAR TIME—SUNDIAL TIME

The sun we actually see is called the *apparent sun*, and the interval between two successive transits of the sun across the same meridian is called an *apparent solar day*. The day commences when the sun is on the observer's inferior or midnight meridian and is reckoned through 24 apparent solar hours to the next lower transit. Therefore, the sun crosses the observer's superior meridian at *apparent noon*, or 12 hours apparent time.

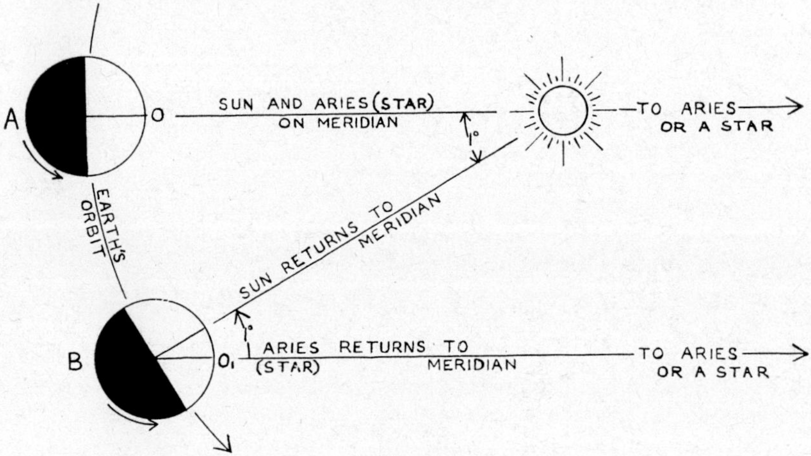

Fig. 4.1
The Sidereal Day is Shorter than the Solar Day.

The sidereal day is shorter than the solar day by about four minutes, as may be verified by noting the interval between successive transits of the sun and any star across the meridian. The reason for the discrepancy between the two kinds of day is the apparent eastwards motion of the sun with respect to the stars due to the earth's orbital motion. This delays the return of the sun to the meridian, making the solar day longer. In fig. 4.1, when the earth is in position A, the observer, O, has the sun and a given star on his meridian at the same instant. It is assumed that a star is a point source of light, far enough distant to consider as parallel the light rays from it reaching the earth. This is somewhat similar to railway lines appearing to diverge from a distant point but which are in fact parallel. This is not so in the case of the sun which is much closer and presents a sizeable disc for us to see.

After the earth has rotated once upon its axis the star is brought back to the meridian to complete the sidereal day interval. In the meantime, however, the earth has moved further forward in its orbit. The 360° of revolution is completed in a year of $365\frac{1}{4}$ days, giving a daily movement of $360/365\frac{1}{4}$°, or nearly 1°. Therefore, after the completion of the sidereal day, the earth must rotate a further 1° in order to bring the sun back to the meridian to complete the solar day, taking nearly four minutes to do so.

Thus, we can choose between taking our time from the stars or from the sun. To use sidereal time for ordinary civil time would not be very sensible because the four minutes daily difference would gradually get us out of step with the sun, leading to a chaotic state of affairs.

The sun is obviously our more natural timekeeper because our daily affairs are more closely governed by it rather than by the stars. Unfortunately, the sun we see is not a good timekeeper for two reasons. Firstly, the earth's orbit around the sun is an ellipse and the motion in its orbit is not uniform. This is reflected as a non-uniform apparent eastwards motion of the sun with respect to the stars which is most rapid when the earth is at perihelion and least rapid at aphelion. Secondly, the sun's ecliptic path is inclined at $23\frac{1}{2}$° to the celestial equator which causes the sun to have a more rapid eastwards motion at the solstices when it is moving parallel to the celestial equator and less rapid at the equinoxes when the sun's path is inclined $23\frac{1}{2}$° to the celestial equator.

These two factors combine to cause irregularities in the length of the apparent solar day such that no ordinary clock would be able to keep pace with the day to day variations—the only instrument capable of keeping time by the sun is a *sundial*. Instructions for constructing such an instrument are given in the practical section. For ordinary civil purposes a day of uniform length is desirable and this has led to the introduction of *mean solar time*.

MEAN SOLAR TIME

For the purpose of defining mean solar time, astronomers assume the existence of an object called the *mean sun* which moves around the celestial equator at a uniform speed, taking exactly the same time to make a complete revolution as the true or apparent sun does to complete its annual journey around the ecliptic. Thus the mean sun gives days each of the same length, *i.e.* equal to the average length of the apparent solar days throughout the year. Although it must be emphasised that the mean sun is an entirely fictitious body, the correct definition of the *mean solar day* is the interval between two successive transits of the mean sun across the same meridian. It commences when the mean sun is on the observer's inferior or midnight meridian and is reckoned through 24 hours to the next lower transit. The day is divided into two equal periods of 12 hours, called A.M. and P.M. (ante-meridian and post-meridian respectively), but in astronomy and other fields a 24 hour notation is adopted, in which case the hours are counted consecutively starting from 0 hours at midnight. Thus 7.20 P.M. becomes 19 h 20 m, or simply 1920 hours.

Each meridian has its own mean time called *Local Mean Time* (LMT) and, since the sun's daily motion is from east to west, it follows that the day begins earlier in places towards the east and later towards the west. Therefore, mean time becomes progressively later the further east one goes and progressively earlier to the west.

During the course of one day the sun crosses every meridian in turn through the full 360° of longitude in 24 hours.
Thus:

360°	of	arc	is	equivalent	to	24 h	of	time
15°	,,	,,	,,	,,	,,	1 h	,,	,,
1°	,,	,,	,,	,,	,,	4 m	,,	,,
1′	,,	,,	,,	,,	,,	4 s	,,	,, .

Example 1

In longitude 30° W the LMT is 08 h 00 m. What is the LMT at the same instant in longitude 60° W?

LMT in 30° W	08 h 00 m
Difference of longitude 30° W	2 h 00 m
LMT in 60° W	06 h 00 m.

The mean time of the Greenwich meridian is called *Greenwich Mean Time* (GMT) or *Universal Time* (UT). It is used all over the world in astronomy, navigation or wherever special activities demand a common time system. Since longitude is measured from Greenwich it follows that at any place the difference between the GMT and the LMT at a given instant is simply the longitude expressed in time. This may be stated as follows:

GMT = LMT + west longitude in time
GMT = LMT — east longitude in time.

A useful way of remembering this is by the rhyme, "longitude west, Greenwich time best—longitude east, Greenwich time least".

Example 2

The LMT aboard a ship in longitude 122° 30′ E is 15 h 20 m. What is the correct GMT?

LMT	15 h 20 m	120°	=	8 h
East long. in time	8 h 10 m	2°	=	8 m
GMT	07 h 10 m	30′	=	2 m
		122° 30′	=	8 h 10 m.

If both the GMT and the LMT are known, the difference between them is the longitude. We shall make use of this knowledge later.

Now that we know what mean time is, the relation between sidereal time and mean time may be clearly stated. The sidereal day is 23 h 56 m 04 s of mean solar time, or 3 m 56 s shorter than the mean solar day. This means that any star will rise, cross the meridian, and set 3 m 56 s earlier each night.

Example 3

A certain star crossed the meridian at 20 h 50 m Civil Time on November 2. What will be the Civil Time of meridian passage on November 22?

Time of transit Nov. 2	20 h 50 m 00 s
20 days at 3 m 56 s earlier	1 h 18 m 40 s
Time of transit Nov. 22	19 h 31 m 20 s.

The reader is left to verify for himself the fact that a star will cross his meridian about two hours earlier each month.

Because of the definite link between sidereal time and mean solar time, astronomers are able to check both against the daily meridian transits of the stars, instead of frantically trying to keep track of a non-existent mean sun.

THE EQUATION OF TIME

During their respective annual journeys, the apparent sun non-uniformly around the ecliptic and the mean sun uniformly around the celestial equator, the apparent sun sometimes forges ahead of the mean sun and sometimes lags behind it. Consequently apparent time may run fast or slow on mean time. At any instant the difference between apparent time and mean time is called the *Equation of Time*. The values of the Equation of Time (EoT) are given in the Sun Table at the end of Chapter 1, at five day intervals throughout the year, as a correction to be applied to Apparent Solar Time (sundial time) in order to give Local Mean Time.

Example 4

On February 4 a sundial in Edinburgh (long. 3° 11′ W) indicated 10 h 40 m. What was the GMT?

Sundial time	10 h 40 m	
EoT	+ 14 m	
LMT	10 h 54 m	
West long. in time	13 m	(actually 12 m 44 s)
GMT	11 h 07 m	

The Civil Time operating in Great Britain is *British Standard Time* (BST) which is one hour ahead of GMT. Therefore, although the sundial shows 10 h 40 m, the correct Civil Time is 12 h 07 m. This considerable difference between the time indicated by a sundial and a watch keeping Civil Time mystifies most people, often resulting in scoffing remarks being made about a sundial because it does not show the "right time".

Examination of the values of the EoT will show that it is zero at four times during the year—in April, June, August and December. The maximum value occurs at the beginning of November when sundial time is nearly 16½ minutes fast of local mean time. In mid-February sundial time is about 14½ minutes slow of local mean time.

ZONE TIME AND STANDARD TIME

We have seen that local mean time becomes progressively later to the east of any place and earlier to the west. If every place kept its own local mean time the inconvenience to travellers would be immense. For example, think of the chaos involved in railway timetables and the like. Imagine leaving on Manchester Time, making a connection on York Time and arriving on Newcastle Time. The situation would be even more chaotic on a world scale, so to minimise the effect of time variation the world is divided into 24 *time zones*, each embracing 15° of longitude. The time kept in each zone is the LMT of the central meridian. Each zone is given a number, prefixed with a plus or minus sign to indicate the number of hours to be added to or subtracted from the *Zone Time* in order to give GMT.

Zone 0 extends between longitudes $7\frac{1}{2}°$ E and $7\frac{1}{2}°$ W and keeps the local mean time of the Greenwich meridian, *i.e.* GMT.

Zone +1 extends between longitudes $7\frac{1}{2}°$ W and $22\frac{1}{2}°$ W and keeps the local mean time of the meridian of 15° W, *i.e.* GMT less one hour.

Zone —1 extends between longitudes $7\frac{1}{2}°$ E and $22\frac{1}{2}°$ E and keeps the local mean time of the meridian of 15° E, *i.e.* GMT plus one hour.

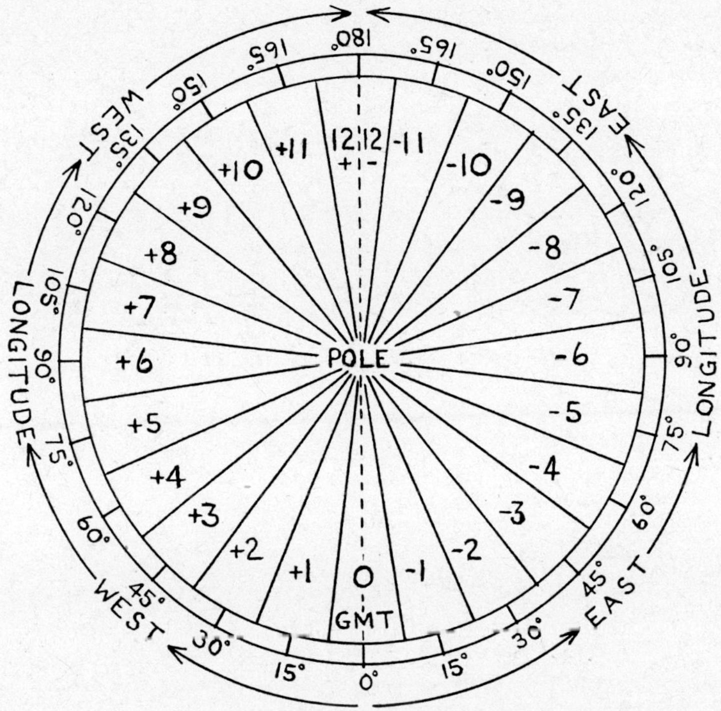

Fig. 4.2
Time Zones around the World.

The time zones are extended progressively eastwards and westwards as shown in fig. 4.2 to zone 12 which extends between longitudes $172\frac{1}{2}°$ E and $172\frac{1}{2}°$ W. The western half of this zone is 12 hours behind Greenwich and the eastern half 12 hours ahead. Therefore, the western half differs from the eastern half by 24 hours, or a whole day. This means that ships and aircraft crossing the 180th meridian from west to east longitude must advance the date by one day. When crossing in the opposite direction the date must be retarded by one day. This makes it possible, on a fast journey, to leave today and arrive yesterday—so the time travellers of science fiction are not unique after all.

The line separating earlier and later dates is called the *International Date Line.* To avoid the division of politically associated areas the date line departs from the 180th meridian in the region of eastern Siberia, the Aleutian Islands and some Pacific Island groups.

The zone time system is followed fairly closely by ships at sea, but on the land the time zones are often modified considerably to satisfy geographical consideration. The time kept in any country is usually fixed by legislation and is known as *Standard Time.* Large countries such as America, Canada and Russia find it necessary to employ a number of standard time zones. Some countries adopt a system of *summer time* or *daylight saving time.* In Great Britain summer time was introduced in 1916; clocks were advanced an hour in the spring of each year and put back in the autumn. In 1968, for an experimental period of three years, it was decided to keep British Standard Time (BST) all year round.

$$\text{BST} = \text{GMT} + \text{one hour}$$

At the end of the period Parliament will decide whether to retain British Standard Time or revert to summer time.*

THE CALENDAR

The functions of a *calendar* are to establish dates of past or future events and facilitate the measurement of time intervals. There are three main types of calendar—the *lunar calendar*, the *lunisolar calendar* and the *solar calendar*.

The lunar calendar is based on the lunar month of 29½ days, *i.e.* the time taken between one new moon and the next new moon. It consists of 12 months, alternately 29 and 30 days in length, making 354 days in all. It is shorter than the solar year, causing the seasons to occur earlier each year. The Mohammedan calendar is of this type.

The lunisolar calendar is an attempt to keep in step both with the phases of the moon and the seasons by adding to the short lunar year a 13th month every two or three years according to fixed rules. The Jewish calendar is of this type.

The solar calendar keeps the length of the year as near as possible to the year of the seasons, the tropical year, which is the

*Great Britain reverted to summer time in 1971.

interval of 365 d 5 h 48 m 46 s between two successive returns of the sun to the Vernal Equinox. Our present calendar is of this type and has no relation to the phases of the moon.

The origin of our calendar dates back to the founding of Rome in the 8th century BC. It was basically a lunar calendar consisting of 304 days divided into 10 months (not all of equal length) named as follows: Martius (March), Aprilis (April), Maius (May), Junius (June), Quintilis, Sextilis, September, October, November and December. Later on, two more months, Januarius (January) and Februarius (February) were added to make a lunisolar type calendar of 355 days in length.

In the 5th century BC the calendar months were rearranged with January at the beginning. This gave the present order of months and accounts for the "number months" not being in their correct order. For example, October is the tenth month and not the eighth, as its name implies.

When Julius Caesar became ruler of Rome in 46 B.C. the calendar was greatly out of phase with the seasons, so upon the advice of the Alexandrian astronomer Sosigenes it was brought up to date by adding 90 days. Because of its undue length, the year 46 B.C. became known as the "year of confusion". In the following year the *Julian calendar* was introduced. This was a solar type calendar similar to that being used by the Egyptians, who recognised the year of the seasons to be 365¼ days in length. To make allowance for the odd ¼ day, three common years of 365 days were to be followed by a fourth of 366 days, called a *leap year*.

After Caesar's death, the month Quintilis was renamed Julius (July). Emperor Augustus gave the months their present lengths and changed the month Sextilis to Augustus (August).

The Julian calendar, based on a year of 365¼ days, was actually 11 m 14 s longer than the true length of the year of the seasons. Consequently the calendar fell behind with respect to the seasons by three days every 400 years. By the year 1582 the Vernal Equinox was occurring on March 11th and would have gone on occurring earlier each year, until Pope Gregory XII took matters in hand. He made two fundamental changes to give us our present calendar, which is known as the *Gregorian calendar*. Firstly, 10 days were dropped from the year 1582 to bring the Vernal Equinox to March 21. The Easter date which depends on this date is regarded as the first Sunday after the first full moon occurring on or after the Vernal Equinox. Secondly, it was decided that three out of four century years were to be *common years* instead of leap years as they would have been under the Julian system. This was achieved by making common years of century years not evenly divisible by 400. Otherwise, leap years would continue to be years evenly divisible by four.

Thus: 1900 was not a leap year (not divisible by 400).
1964 was a leap year (divisible by four).
1966 was not a leap year (not divisible by four).
1968 was a leap year (divisible by four).
2000 will be a leap year (divisible by 400).

The Gregorian calendar is now used by most countries of the world. Great Britain and the USA (then a British Colony) adopted it in 1752 and the USSR in 1918.

On average, the Gregorian calendar is still 26 seconds longer than the year of the seasons but this deficiency will not merit attention for another few thousand years.

REVIEW QUESTIONS

1. Name the point or object in the heavens associated with the measurement of sidereal time, apparent time and mean time.

2. Explain why the sidereal day is 3 m 56 s shorter than the solar day.

If a star crosses an observer's meridian at 22 h 00 m tonight, at what time will it transit his meridian two weeks later?

3. An Apollo spacecraft is launched from Cape Kennedy at 11 A.M. local time and live transmissions via satellite are seen on British television at 4 P.M. local time. Explain clearly the reason for the time difference.

4. A ship is in longitude 35° 15′ E and the ship's chronometer, which keeps GMT, indicates 10 h 40 m. What is the correct local mean time of the meridian on which the ship is situated? If the ship's clocks are keeping Zone Time, what time do they indicate?

5. What term is used to describe the difference between apparent and mean solar time?

On November 11, a sundial at Oxford (long. 1° 15′ W) shows 12 noon. What is the British Standard Time to the nearest minute?

6. A ship's officer comes off duty at 0800 hours on Sunday when the ship, steaming west, is about to cross the date line. What will be the time and day when he resumes duty eight hours later?

7. State the main reforms introduced into the Julian and Gregorian calendars. Which of the years 1980, 1992 and 2000 are leap years?

8. A person's birthday occurs on Wednesday in March in a year following a leap year. On what days will his birthday fall in the three succeeding years?

PROJECTS AND RESEARCH

1. MAKING A STAR CLOCK

The constellation of The Plough is circumpolar as seen from any place in the British Isles, in fact from any place whose latitude is greater than 41° N. This means that the stars of The Plough appear to trace out daily circular paths around the Pole Star, providing us with a form of clock in the sky. This sky clock is different from an ordinary clock in as much as the movement is anticlockwise and one revolution takes a whole day. Another important difference arises from the sidereal day being shorter than the mean solar day. This causes the stars to return to the same position in the sky four minutes earlier each night or two hours earlier each month, but can be allowed for by making our clock dial capable of being set back two hours each month.

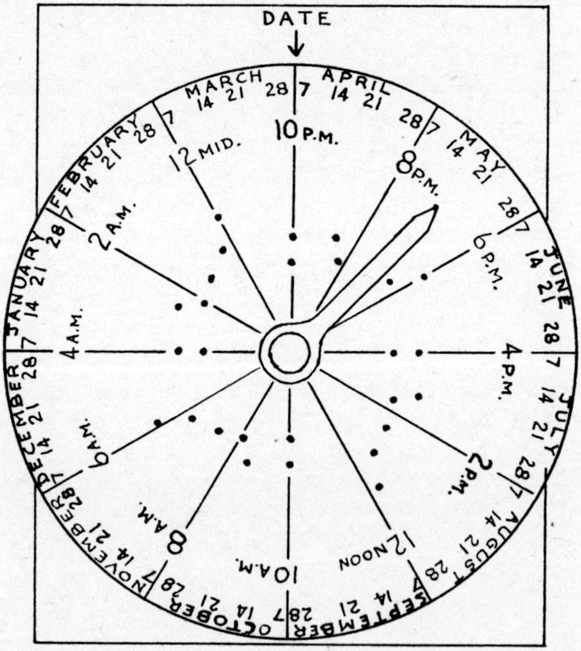

Fig. 4.3
Construction of a Star Clock.

A star clock similar to the one illustrated in fig. 4.3 can be constructed as follows:

First, obtain a thin, flat piece of wood, measuring 5 in by 4 in, (127 mm by 102 mm) to act as a solid baseboard for the moveable clock face. Then, from a piece of stiff card cut out a circular disc of $4\frac{1}{2}$ in (114 mm) diameter and with the aid of a protractor draw lines from the centre point of the disc 30° apart. On the disc draw circles of radii $2\frac{1}{8}$, 2 and $1\frac{3}{4}$ in (54, 51 and 45 mm). Between the edge of the disc and the outer circle insert the names of the months, and between the two outer circles the dates 7, 14, 21 and 28 in a clockwise sense, with the 7 against the radial lines. On the inner circle insert local mean times starting with 12 midnight on March 7; followed by 10 P.M. on April 7; 8 P.M. on May 7 and so on back to midnight again. Next, out of stiff card, make a pointer of sufficient length for it to be rotated just within the figures of the time scale. Position the disc on the baseboard such that it overlaps $\frac{1}{4}$ in (6 mm) on each side (this makes it easier to rotate the disc) and then push a drawing pin through the pointer and the centre of the disc into the baseboard. At the top of the baseboard, directly above the centre of the disc, make a clear mark to serve as a date indicator. This completes the construction, although the disc may be decorated with several outlines of The Plough, as shown in the figure.

To use the star clock, rotate the disc until the nearest date coincides with the date indicator. Hold the clock vertically, with the thumb pressing the disc to the baseboard to prevent movement, and direct it upwards until it is centred on the Pole Star. With the other hand move the pointer so that it is in line with the "pointers" of The Plough. The local mean time is then indicated by the pointer on the time scale. In the British Isles this may be taken as GMT (ignoring slight differences of longitude from place to place) but do not forget to add one hour for BST. With care and practice it should be possible to obtain the time within a quarter of an hour.

2. MAKING A SUNDIAL

A sundial is an instrument designed to keep local apparent time, *i.e.* time according to the sun we see. It consists basically of a dial graduated in hours, over which moves the shadow cast by a device called a gnomon or style.

It may be thought that a circular dial graduated in divisions of 15° to one hour and having a vertical style placed at its centre would make an admirable sundial, but unfortunately it would not; try it and see. Even the early Greek and Arab astronomers realised that a dial graduated in this way would give a correct indication of time only if the instrument was tilted so that the style was elevated to the latitude of the place. This made the style parallel to the earth's axis, and the plane of the dial parallel to the celestial equator and hence parallel to the plane of the sun's apparent daily path across the sky.

A sundial made in this manner is simple enough to make but an interesting and more common variation of construction involves making a horizontal dial. The style is still elevated to the latitude of the place but, because the dial has been returned to the horizontal position, an adjustment must be made to the hour divisions if instrument is to keep correct time by the sun. To get the correct hour spacings involves a certain amount of calculation, which may be unfamiliar to some readers, therefore, the method outlined below is suggested. The measurements given are correct for latitude 54° N but may be assumed correct for any place in Great Britain without incurring more than a few minutes error (see fig. 4.4).

Obtain a smooth, flat wooden board measuring 8 in by 8 in (204 mm sq). Draw a central line, called the gnomon line, down the board from top to bottom and another line, called the base line, across the board at right angles to the gnomon line 2 in (51 mm) from the bottom edge of the board. Along the top edge of the board mark off points $1\frac{1}{4}$ in (32 mm) and $2\frac{3}{4}$ in (70 mm) on each side of the gnomon line. From the base line mark off a point on each side of the board $1\frac{5}{16}$, $2\frac{7}{8}$ and $4\frac{15}{16}$ in (33, 73 and 126 mm) above the baseline, and also one point on each side $1\frac{5}{16}$ in (33 mm) below the baseline.

Draw lines from each point to the point of intersection of the gnomon line and the baseline. These are the hour lines. The gnomon line is numbered 12 noon, the lines to be left in decreasing order of hours to 5 A.M. and the lines to the right in increasing order to 7 P.M.

The gnomon may be cut from a piece of stiff cardboard or other thin, rigid material, to the form of a right angled triangle having a 6-in base and a base angle of 54°. The gnomon is then set up vertically along the gnomon line, with the point of the 54° angle centred on the intersection of the gnomon line and the base line, and glued into position. The gnomon line may be scored or cut out to help in securing the gnomon vertically. The gnomon need not be right angled as stated earlier provided the upper slant edge is 54° to the horizontal.

The completed sundial is then laid on a flat surface at a spot

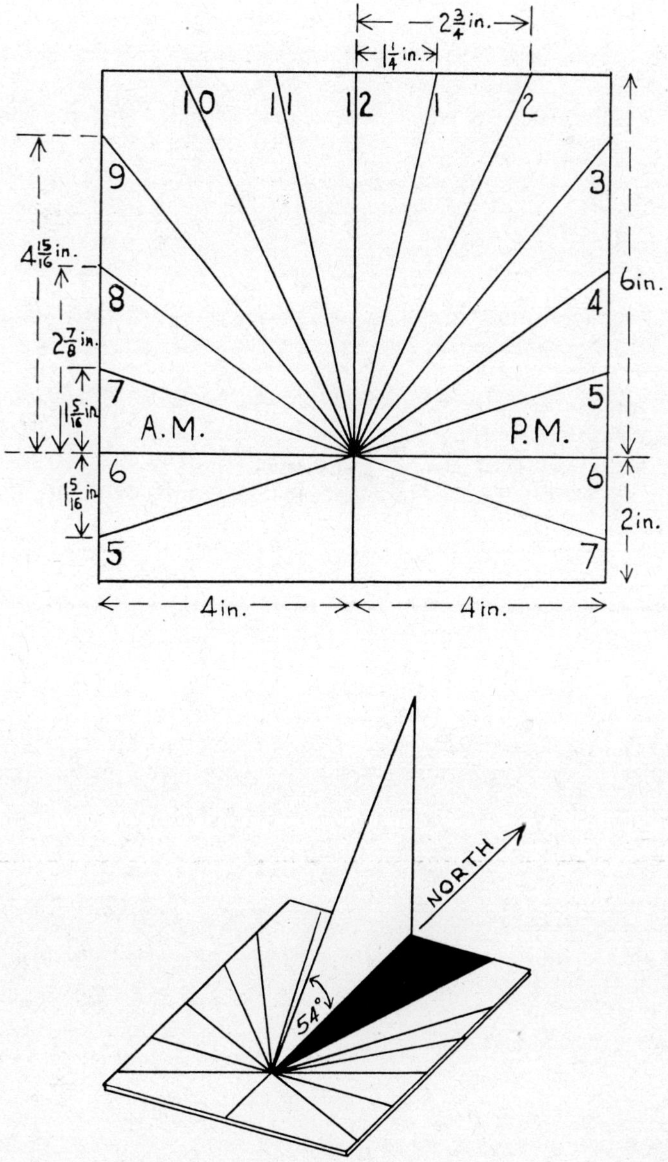

Fig. 4.4
Construction of a Sundial.

where it will receive maximum sunshine, the gnomon line being set in the meridian with the 12 noon mark pointing due north. The sundial is then ready for use, but remember that the time indicated by the shadow on the dial must be corrected as described in Example 4, in order to obtain *Civil Time*.

A knowledge of trigonometry allows the dial to be graduated exactly for any latitude by means of the formula:

$$\tan A = \sin L \times \tan H$$

where A is the angle of any given hour mark from noon, L is the latitude and H the angular distance of the sun from the meridian before or after noon (called hour angle). Hour angle increases 15° for each hour from noon. Therefore, at 11 A.M. and 1 P.M. sundial time H is 15°, at 10 A.M. and 2 P.M. H is 30° and so on. The construction for a southern hemisphere location is similar to that for the northern hemisphere, except that the A.M. hour marks are on the right and the P.M. marks on the left. The gnomon line is set in the meridian with the 12 noon mark pointing due south.

3. FINDING LONGITUDE

Suppose a meridian observation of the sun is being made for the purpose of finding latitude, as described in Project 3, Chapter 1. Then, with the aid of a reliable watch, corrected prior to the observation by means of a radio time signal or the Post Office telephone speaking clock, it is possible for the observer to find his longitude as well as his latitude.

At the exact instant the sun reaches its maximum altitude, make a note of the watch time, if possible to the nearest $\frac{1}{10}$ minute. In Great Britain, the watch time will be BST so it is necessary to subtract one hour in order to obtain GMT.

It will be remembered that the sun is on the meridian at 12 h 00 m sundial time or, more correctly, local apparent time (LAT). The Equation of Time, extracted from the Sun Table, and applied to 12 h 00 m will provide the LMT of observation.

The difference between the LMT and the GMT, converted into degrees, is the longitude.

Example 5.

On July 25, the altitude of the sun at upper transit, bearing due south, was 54°·5. The correct watch time of observation was 13 h 12·2 m. Calculate the latitude and longitude of the observer.

	90°·0	
altitude	54°·5 south	
zenith distance	35°·5 north	like names—add
declination	19°·5 north	
latitude	55°·0 north	

LAT	12 h 00·0 m	BST	13 h 12·2 m
EoT	+ 6·4 m	—	1 h 00·0 m
LMT	12 h 06·4 m	GMT	12 h 12·2 m
		LMT	12 h 06·4 m
		Long. in time	5·8 m

4 m = 1°; 1·8 m = 0·45°, hence 5·8 m = 1°·45 or 1°27′.
Remember—longitude west, Greenwich time best.
Therefore, observer's position is: lat. 55° 00′ N
long. 1° 27′ W

Alternatively, the time of maximum altitude, which coincides with the time of shortest shadows, may be obtained from shadow stick measurements. Suppose the watch times of equal morning and afternoon shadows (see Project 1, Chapter 1) were respectively 11 h 00 m and 15 h 24·4 m. The watch time (BST) of meridian passage is midway between these times, *i.e.* 13 h 12·2 m, and the procedure for calculating longitude is then similar to that shown above.

4. FINDING THE EQUATION OF TIME

We have seen that a knowledge of the Equation of Time is necessary for use with the sundial. Although its value is tabulated in the Sun Table there is no reason for these figures to be accepted, as it is not difficult to determine the Equation of Time from practical observations. In this case, however, it is necessary to know the longitude of the location from which observations are made.

The watch time of the meridian transit of the sun is noted and the corresponding LMT calculated. This is then compared with the LAT (12 h 00 m) of the observation.

Example 6.

At Great Yarmouth (long. 1° 45′ E) the watch time of the sun's meridian passage on June 25 was 12 h 55·5 m. Find the Equation of Time for that day.

BST		12 h 55·5 m	
	—	1 h 00·0 m	
GMT		11 h 55·5 m	
E long. in time	+	7·0 m	(long. E, Greenwich time least)
LMT		12 h 02·5 m	
LAT		12 h 00·0 m	
EoT	+	2·5 m	

The value of the Equation of Time is given a plus sign because, for our purposes, we are regarding it in the sense of a correction to be applied to sundial time in order to give mean time and thence Civil Time.

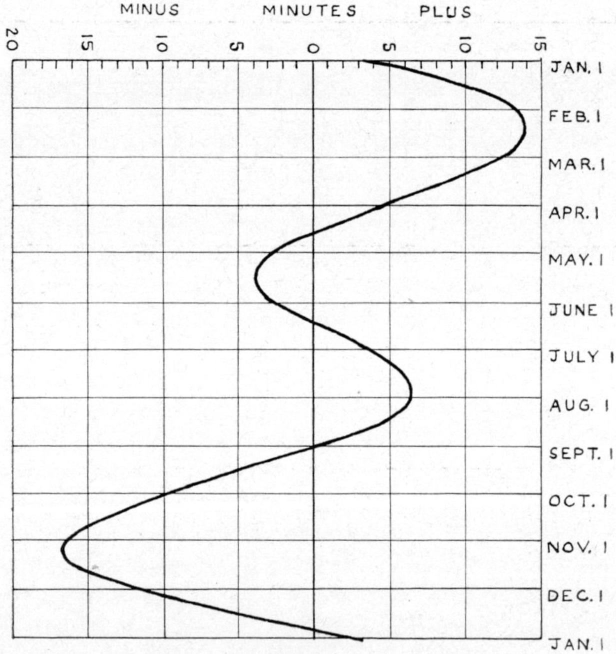

Fig. 4.5
The Equation of Time.

If observations are carried out at intervals of several days throughout the year it is possible to make a graph of the Equation of Time similar to the one shown in fig. 4.5.

5. SHADOWING THE SUN

This project involves plotting the position of the sun's shadow at regular intervals throughout the year in order to demonstrate the difference between mean time and apparent time. A 6-in (15 cm) vertical stick is mounted on a horizontal baseboard on which is pinned a sheet of drawing paper of sufficient size to accommodate the shadow. If observations are to be made around midday, some

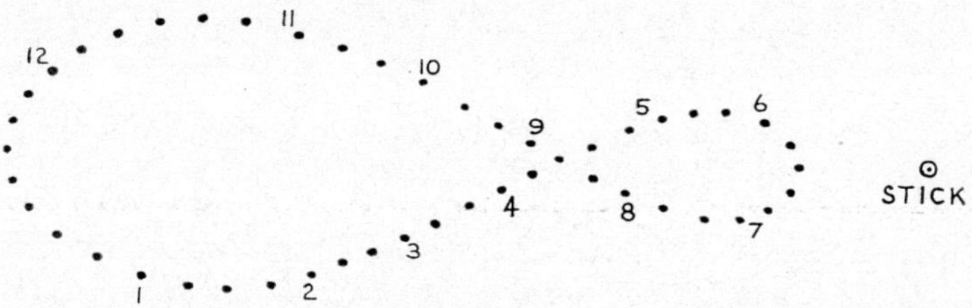

Fig. 4.6
Shadowing the Sun.

idea of the size of the paper required can be obtained by calculating the length of the longest shadow, which occurs at the winter solstice. The altitude is obtained by the methods described in Project 2, Chapter 2, and the length of the shadow by scale drawing. The equipment must be set up in the same position at each observation, so it would be preferable if it could be fixed permanently in a sheltered location. At intervals of several days, mark the extremity of the shadow at the instant of the noon time signal broadcast by the national or local radio station. As the year progresses, it will be seen that the pattern traced out by the shadow looks like a "figure eight", with a very elongated autumn-winter loop and a smaller spring-summer loop. The cross-over point of the loops and their extremities coincide with the four dates when the Equation of Time is zero. A line drawn through these points represents the datum line, and the displacement of any point on the shadow from it is the value of the EoT. It would, of course, be necessary to obtain a few values of the EoT, either by experiment as described in the previous project or from the Sun Table in Chapter 1, in order to establish the scale of the shadow curve. The figure eight diagram obtained by shadowing the sun is known as an *analemma*.

ANSWERS TO REVIEW QUESTIONS

2. 21 h 04 m 56 s.
4. 13 h 01 m LMT, 12 h 40 m ZT.
5. 12 h 49 m BST.
6. Monday 1600 hours.
7. All are leap years.
8. Thursday, Friday, Sunday.

1. Polar Star Trails — *2. Star Trails in Orion*

(Left) the bright star close to the centre is Polaris. (Right) Orion is clearly distinguishable. The bright trail, bottom left, is Sirius.

Photographs by GORDON PERCIVAL, *with a fixed camera. Each exposure 30 minutes.*

PLATE I

3. Aurora: Rayed Band
Photographed by MR. BEEDEN *from a location in Iceland.*

4. The Full Moon
Photograph taken by ASHLEY ROSS, *through the eyepiece of a 2-in refractor. Exposure $\frac{1}{3}$ second.*

PLATE II

CHAPTER 5

THE MOON

In its annual journey around the sun the earth is accompanied by its natural satellite, the moon; a rugged barren world of 2160 miles (3476 km) in diameter. Of the 32 known satellites associated with the planets of the solar system the moon ranks sixth in order of size, but a comparison between the sizes of satellites and parent planets shows that relatively speaking the moon, with a diameter of just over a quarter that of the earth, is so large as to give the earth and moon together the appearance of a "double planet."

THE MOON'S ORBIT AND PHASES

The earth and moon are attracted to one another by the force of their mutual gravitation, and actually revolve around each other about their common centre of mass, situated approximately one quarter of the earth's radius below the earth's surface. The situation is not unlike that of heavyweight and lightweight boxers sparring in the ring, the heavyweight being thought of as the earth; the lightweight the moon. In this unequal contest the "earth" moves in a tight circle, whereas the "moon" prances around in a much wider arc, doing most of the work. In the earth-moon system the motion of each body is further complicated because they also revolve around the sun. In fact, it is the centre of mass which performs the elliptical orbit around the sun, whilst the earth and moon weave in and out of it. For the present we shall consider the moon's motion relative to the earth's centre, saying simply that it revolves around the earth in a slightly elliptical orbit, with the earth at one focus of the ellipse. The point in the orbit where the moon is nearest to the earth is called *perigee* and the point where it is furthest away is called *apogee*; its mean distance from the earth being 239 000 miles (384 544 km). When viewed from the earth, the moon appears slightly smaller at apogee than it does at perigee but its mean angular diameter is about $\frac{1}{2}°$—the first finger and thumb held $\frac{1}{5}$ in (5 mm) apart at arms length will just embrace the full moon.

The moon is a solid body which shines by reflection of sunlight, half the lunar globe being illuminated at all times. As the moon revolves around the earth so the illuminated hemisphere is presented to us in varying amounts, thereby giving rise to the changing phases we see (fig. 5.1).

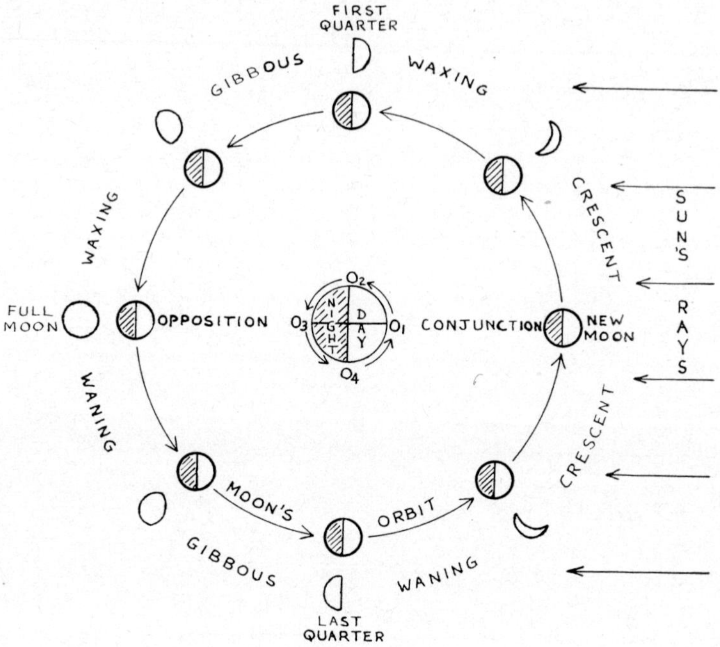

Fig. 5.1
Phases of the Moon.

New moon occurs when the moon passes between the sun and the earth. At this time the dark hemisphere is turned towards us and the moon is invisible unless it happens to pass directly in front of the sun, in which case an eclipse of the sun occurs. A day or so after new moon a slender crescent is likely to be seen in the west at sunset. Most people call this the new moon because it is the first sighting of the moon after it disappeared on approaching the general direction of the sun towards the end of the previous succession of phases. About seven days after new phase the moon has performed one quarter of its revolution and the gradually thickening crescent develops into a half phase, known as *first quarter*. The line separating the dark and bright hemispheres of the moon, the *terminator*, appears to swell outwards after first quarter to give the *gibbous phase*, which increases night after night until *full moon* is reached. This occurs at just under 15 days after new moon and the moon is then on the opposite side of the earth to the sun.

While the moon appears to be growing it is said to be *waxing*, and the terminator represents the line of sunrise on the face of the moon. After full phase the terminator represents the line of sunrise on the face of the moon. After full phase the terminator represents the line of sunset, and as it creeps over the lunar surface the phases are repeated in reverse order. While the moon appears to be shrinking it is said to be *waning*, and about five days after third, or *last quarter* the moon becomes drowned in the strong light of the sun.

Again referring to fig. 5.1, an observer on the rotating earth at 0_1 has both the sun and the new moon on his meridian at midday. At first quarter, when the observer is at 0_1, the sun is on the meridian but the moon will not be on the meridian until the observer has been carried around through a quarter of the earth's rotation to 0_2, taking a further six hours. Therefore, at first quarter the moon crosses the meridian at about 6 P.M. local time. By similar reasoning, after the sun is on the meridian, the observer has to be carried around through a half rotation of the earth from 0_1 to 0_3 before the full moon is on the meridian, and three quarters of a rotation to 0_4 before last quarter is on the meridian. Thus, full moon is on the meridian about midnight local time; last quarter at 6 A.M.

As seen in the lunar sky the earth passes through a succession of phases too, the phase of the earth at any time being opposite to the phase of the moon seen from earth. The earth is a better reflector of sunlight than the moon and, of course, the earth is larger than the moon, so full *earthlight* on the moon is much brighter than the light of full moon on earth. When the moon is in crescent phase the earth is in gibbous phase and the stronger earthlight reaching the moon is partially reflected back to earth. On a clear night, at the time of crescent moon, not only can we see the crescent illuminated by the sunshine, but also the rest of the lunar disc dimly illuminated by *earth shine*. The horns, or cusps, of the crescent moon seem to be wrapped around the earthlit part, giving rise to the country saying "the old moon is the young moon's arms".

THE LUNAR MONTH

The true period of the moon's revolution around the earth, called the *sidereal month*, is the time taken for the moon to return to the same direction in space. The average length of the sidereal month is $27\frac{1}{3}$ days.

We saw earlier that the sun has an eastwards motion with respect to the stars. The moon has a similar motion but, because it is very much closer to the earth than the sun, its eastward movement against the starry background is much more rapid. Considering the stars as fixed points in space, the sidereal month can be regarded as the interval between two successive returns of the moon to the same star.

The interval between two successive new moons, known as the *synodic month*, or a *lunation*, has an average length of 29½ days.

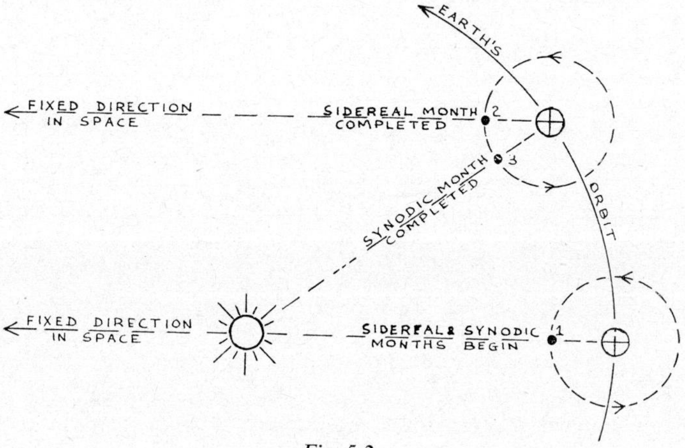

Fig. 5.2
The Sidereal Month is Shorter than the Synodic Month.

The reason for the difference in length of the sidereal and synodic months is illustrated in fig. 5.2. When the moon, in position 1, is between the earth and sun, it is new moon. About 27⅓ days later the moon, in position 2, has returned to the same direction in space, thus completing the sidereal month. In the meantime, however, the earth has advanced some way in its orbit around the sun, the next new moon being delayed by a little over two days until the moon has reached position 3, when it is again between the earth and sun.

THE MOON RISES LATER EACH DAY

The eastward movement of the moon against the background of the stars during the course of the sidereal month amounts to 360° in 27⅓ days. A little arithmetic shows the daily motion to be about 13°, and the hourly motion a little more than ½°. If, on a clear night, the moon is observed on the meridian when it is near to a bright star the difference in their respective motions can be readily seen. Both the moon and the star traverse across the sky towards the west due to the rotation of the earth, but after about an hour the moon will have fallen behind the star by about $\frac{1}{2}°$, or a little more than its own diameter. On the following night when the star returns to the meridian the moon will be lagging well behind the star and will not cross the meridian until over 50 minutes later.

The time of rising of the moon is also retarded by about 50 minutes on average, although the actual retardation can vary considerably, depending on how the moon's path lies with respect to the horizon. In Great Britain the times of moonrise on successive nights can vary by as little as several minutes or as much as $1\frac{1}{2}$ hours. The least delay is most noticeable in the case of the full moon occurring near the Autumnal Equinox (September 23). At this time of year, the moon's path being inclined at only a small angle to the horizon, the full moon rises in the early evening with only a few minutes delay on successive nights. Because its beneficial light enabled farmers to work later in the fields, the autumn full moon became known as the "harvest moon".

THE MOON'S APPARENT PATH

The apparent path followed by the sun around the celestial sphere during the year is the ecliptic.

The moon's path on the celestial sphere during the course of a

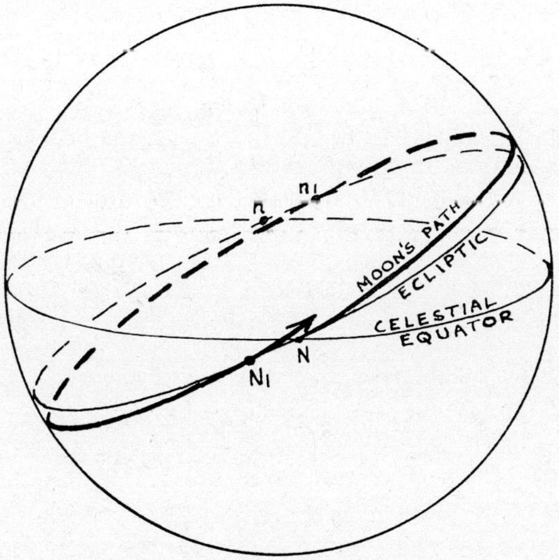

Fig. 5.3
Regression of the Moon's Nodes.

month is nearly a great circle, inclined at about 5° to the ecliptic. Thus, in a year the moon makes $12\frac{1}{2}$ journeys around the celestial sphere, whilst the sun makes only one. The points where the moon's path crosses the ecliptic are called the *nodes*, the *ascending node* being the point where the moon crosses going north; the *descending node* where it crosses going south.

Although the inclination of the moon's path to the ecliptic remains about 5°, the path itself gradually slides along the ecliptic such that each time the moon returns to the ecliptic it is a little further to the west, as shown in fig. 5.3, in which the ascending and descending nodes have been displaced respectively, from N to N_1 and n to n_1. This displacement, called *regression of the nodes*, is continuous and it takes about $18\frac{1}{2}$ years for the nodes to make one complete movement around the ecliptic. In fig. 5.4a the ascending

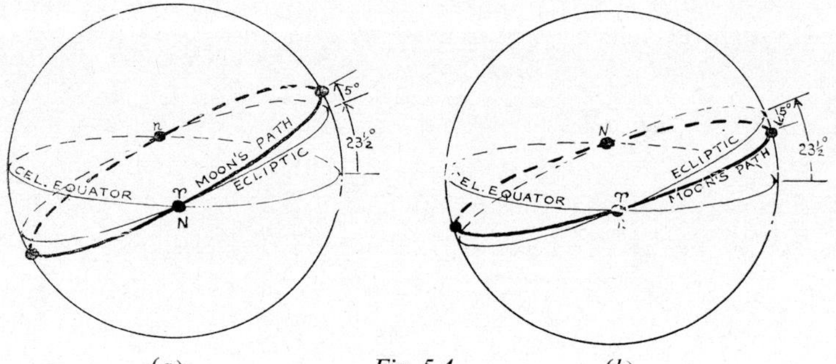

(*a*) *Fig. 5.4* (*b*)
Range of the Moon's Declination.

node N is shown coinciding with the First Point of Aries (♈) at the time of the Spring or Vernal Equinox. Just over nine years later the moon's path has slewed half way round the ecliptic so that the ascending node occurs at the Autumnal Equinox (fig. 5.4b). This unusual displacement of the moon's path can be more easily understood by bending two lengths of wire into circles and moving one around inside the other, keeping the angle between them constant. The effect of regression is to cause a wide variation in the declination of the moon. When the ascending node coincides with the Spring Equinox (this occurred in 1969) the maximum angular distance of the moon from the celestial equator amounts to $23\frac{1}{2}°$ plus 5°, or $28\frac{1}{2}°$. When the ascending node coincides with the Autumnal Equinox (occurs in 1978) the maximum angular distance of the moon from the celestial equator is $23\frac{1}{2}°$ minus 5°, or $18\frac{1}{2}°$. In any month, therefore, the moon's declination can range from as much as $28\frac{1}{2}°$ N to $28\frac{1}{2}°$ S or from as little as $18\frac{1}{2}°$ N to $18\frac{1}{2}°$ S depending on where the nodes of the moon's path are on the ecliptic during the $18\frac{1}{2}$ year regression cycle.

The rapid change in the declination of the moon is accompanied by a noticeable variation in its altitude at meridian passage. Take the case of the full moon which is always near the opposite point on the celestial sphere to the sun. In northern summer, when the sun has north declination, the full moon has south declination and hangs low in the sky. In northern winter, when the sun has south declination, the full moon has north declination and rides high in the sky, as though nature intended it to brighten the long cold nights. Of course the reverse is true in the southern hemisphere. The reader is invited to figure out for himself the times of year when the young moon and quarter moon attain maximum altitude.

THE MOON'S ROTATION AND LIBRATIONS

Earthbound observers can never view the far side of the moon directly for, as it revolves around the earth, our satellite always keeps the same face turned towards us. This often leads to the mistaken belief that the moon does not rotate, but in fact it does spin once on its axis in exactly the same time as it takes to complete one revolution. The truth of this statement can be verified by the following simple experiment. Stand in a room directly facing one of the walls and then turn right around on your heels so that you face each wall in order after each successive quarter turn—this is rotation. Now move around in a circle as though you are the moon orbiting the earth, which is assumed to be at the centre of the circle—this is revolution. If you keep your face turned towards the "earth" you will also face each wall in the same order as you did when performing rotation only. After each additional quarter revolution you will also have completed a further quarter rotation, thus keeping the rotation and revolution in step with each other. The moon behaves in the same way with respect to the earth, and this is why we always see the changing face of the "man in the moon"—never the back of his head.

From what has been said in the preceding paragraphs it may be thought that only half of the moon's surface can ever be seen from the earth. Yet long before spacecraft went anywhere near the moon about 59 per cent of its surface had already been mapped by astronomers. There are several reasons why this is possible, although we shall only concern ourselves with the two most important of them. Firstly, in the same way as the earth moves in its orbit with its axis tilted, so the moon moves in its path with its axis inclined about $6\frac{1}{2}°$ to the perpendicular to the plane of its orbit. This means that the north pole of the moon is tipped towards us for half the month and the south pole for the remainder, enabling us to see a little of the far side around the polar regions. Secondly, the moon's rotation does not keep exactly in step with the revolution as indicated earlier. Although they both start and finish together, the rotation takes place at a uniform rate whilst the revolution varies because the moon travels in an elliptical orbit. The effect of this is to cause the moon to appear to rock slightly on its axis in an east-west direction, so that we get a glimpse around each side in turn in the course of a lunation.

These apparent movements of the moon seen from earth are called *librations*. The extra view we get due to the librations in latitude and longitude, described above, together with other smaller librations, leaves only 41 per cent invisible from earth. The far side of the moon was first photographed from the Russian spacecraft Lunik 3 in 1959 but since then there have been numerous successful lunar probes, both by the Russians and Americans, and to date most of the moon's surface has been scanned in close-up. In 1968, three American astronauts, Borman, Lovell and Anders, took their Apollo 8 spacecraft around the moon to become the first men ever to view its far side directly.

SURFACE CONDITIONS ON THE MOON

When viewed with the naked eye the surface of the moon appears to be composed of extensive dark and bright areas, which together form the characteristic face of the man in the moon.

The dark areas are the lunar *maria* (seas), so called because early astronomers thought them to be large expanses of water. They were given such fanciful names as Mare Tranquillitatis (Sea of Tranquility), Mare Serenitatis (Sea of Serenity) and Oceanus Procellarum (Ocean of Storms). Later, with the development of larger and better telescopes, the moon was found to be devoid of water but the old names have been retained. The maria, in fact, resemble vast plains believed by many astronomers to be the result of large outflows of lava-like material which spread over much of the moon's surface and then solidified. Although the maria are generally smooth compared with the rest of the moon, there are considerable variations in level together with numerous small craters, low gentle swellings (domes), ridges and long, narrow, winding valleys (rilles). It was once thought that the maria were covered with dust to a depth of many feet, but the soft landing of spacecraft upon the moon has shown the surface to be quite firm. During 1969, the year of Apollo, men walked upon the moon, leaving their foot-prints in the light dust of the Mare Tranquillitatis and the Oceanus Procellarum.

The bright areas of the moon are extremely rugged, comprising a vast complex of mountains and high-walled craters. The mountain ranges are somewhat similar to those on earth, some having familiar names like the Alps, Apennines and Caucasus. Heights on the moon are difficult to compare because there is no sea level as on earth, but the highest peaks are estimated to rise to about 26 000 ft (7900 m) above the level of the neighbouring plains.

The craters form probably the most interesting and puzzling of the moon's surface features. There are literally thousands upon thousands of them, mostly circular in shape, ranging in size from minute holes to great walled plains exceeding 150 miles (241 km) in diameter. Generally speaking, the outer walls rise gently from the surrounding levels whilst the inner walls fall more steeply to the crater floor, sometimes in the form of steps or terraces. In some cases the heights of the crater rim is well in excess of 20 000 ft (6100 m) above the floor. The larger ones have fairly flat floors, interrupted in places by smaller craters and rilles and, in many instances, central mountains. Most of the larger craters are named after persons who have contributed in some way to the advancement of astronomy. The craters Tycho and Copernicus, in particular, possess bright ray systems which can be seen to advantage with a pair of binoculars or a small telescope at around the time of full moon.

The word "craters", used to describe the ring formations on the moon, does not mean they are volcanic features. Perhaps some are extinct volcanoes, but the majority bear little resemblance to those found on earth. Their origin is still open to discussion among astronomers, some believing them to be the result of internal upheavals during the early stages of the moon's development, others regarding them as impact craters caused by the collision of meteorites with the moon at a time when such bodies were more numerous in interplanetary space than they are now.

Unlike the earth, the moon has no atmosphere or water so there can be no weather as we know it—no wind, rain, snow or ice. Whilst the earth's surface features are undergoing a continuous process of change due to the eroding forces of weather and water, the moon's surface is probably much the same as it was after solidification took place millions of years ago. There have been a few reports in recent years of minor changes on the surface, though these have done nothing to significantly alter the view that the moon is a dead world. Its surface is subjected to extremes of heat and cold, since the lunar days and nights are about 14 times as long as ours. The absence of an atmosphere means that all the dangerous radiation emanating from the sun, from which our atmosphere thankfully protects us, beats down relentlessly upon the moon and in the blistering heat of the sun's rays the temperature can rise to 30 centigrade degrees above the boiling point of water. Then, during the long lunar night, with no atmosphere or cloud cover to prevent heat escaping into space, the temperature can fall to 140 centigrade degrees below freezing point. Another distinctive feature of the moon is that its surface gravity is only about one-sixth of the earth's. Therefore, a lunarnaut who weighs 150 lb (68 kg) on earth weighs only 25 lb (11·3 kg) on the moon. Investigations into the lunar rock samples brought back by Apollo astronauts show that the moon is totally different in structure to the earth.

MAN-MADE MOONS

The space age began on October 4, 1957, when the Russians placed the first artificial satellite, Sputnik 1, into orbit around the earth. Nowadays, the launching of a satellite attracts hardly any general attention; even men travelling to the moon have to compete with other items of world news for a place on the "front page". At the present time there are hundreds of man-made objects circling the earth. Many of them are just "space junk", but others perform specific tasks and send back to earth valuable information which it is impossible to obtain from earthbound observatories.

To launch a satellite into orbit around the earth requires an enormous amount of technological skill. The *circular orbit* is the most difficult to achieve because after the rocket has lifted the satellite vertically upwards through the earth's atmosphere it must be levelled off at the right height and projected horizontally at the correct speed. The early stages of the rocket fall to earth after their fuel has been expended but the last stage, which provides the final thrust, goes into orbit with the satellite after separation has taken place. A satellite is kept in orbit by the force of gravitation: it is attracted towards the earth's centre, so that like the moon it "falls" continuously around the earth. Since gravity decreases with increase of height, there is for every height a correct speed for a satellite to balance the earth's gravitational attraction. Should this speed not be acquired, or if it is exceeded, the satellite will fall towards the earth or go into an *elliptical orbit*. If its speed exceeds 25 000 mph (40 000 km/h), *escape velocity*, it will break away from the earth's gravitational influence and go into orbit around the sun as an *artificial planet*. The speed necessary for a circular orbit at a height of 300 miles (480 km) is about 18 000 mph (29 000 km/h), but even if the speed is correct, any departure from the horizontal at the point where the satellite is ejected from the last stage of the rocket will result either in its plunging earthwards or going into an elliptical orbit. Needless to say, the majority of orbits are elliptical and, as with the moon, the nearest point to the earth is called perigee and the furthest point apogee.

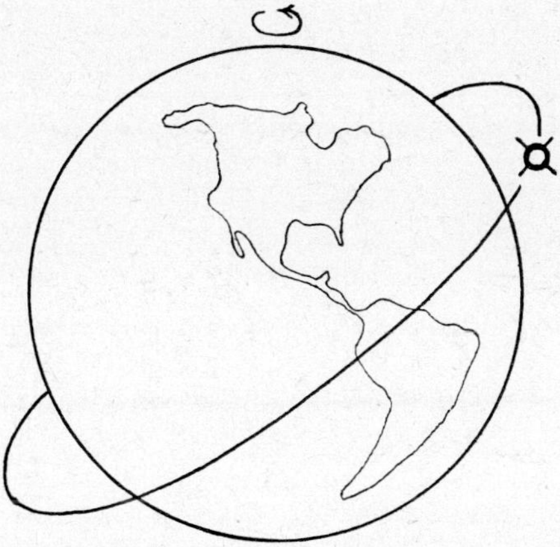

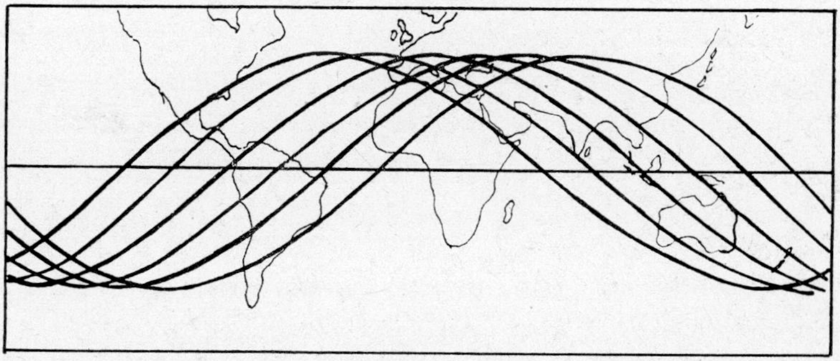

Fig. 5.5
Westward Displacement of a Satellite's Ground Track.

If the satellite traverses over the earth's poles it is said to be in a *polar orbit*; if it is carried around over the equator it is in an *equatorial orbit*. All other orbits are called *intermediate*.

For the moment it is convenient to think of the plane of a satellite's orbit as being fixed in space, with the earth rotating inside it. Since the inclination of the orbit to the equator tends to remain constant, the satellite moves over the earth within fixed limits. For example, if the inclination is 48°, the ground track is bounded by the parallels of 48° N and S. If the period of revolution is 90 minutes, then each time the satellite completes one revolution the earth has performed $\frac{1}{16}$th of one rotation, so that after each orbit the ground track is displaced $22\frac{1}{2}°$ to the west (fig. 5.5). In practice, the orbit is not fixed in space but is modified slowly. Firstly, the greater attraction of the earth's equatorial bulge, and the gravitational pull of the moon and the sun, slews the orbit around by a small amount each day. Secondly, a satellite within the limits of the earth's atmosphere is retarded by air drag, which causes the orbit to shrink gradually until the satellite falls towards the earth and burns up in the heat generated by friction with the thicker atmospheric layers.

Artificial satellites serve a variety of purposes, some of which are given below.

(1) Measurement of the earth's shape and size.

(2) Geodetic and geological surveys.

(3) Investigation into the composition and density of the earth's atmosphere at different heights. Sounding rockets are also used for this purpose.

(4) Measurement of the strength and extent of the earth's magnetosphere and interplanetary magnetic fields.

(5) Investigation of the solar wind and its effect on the magnetosphere, ionosphere, and the Van Allen radiation belts.

(6) The provision of intercontinental radio, telephone and television links—Echo, Telstar, Relay, Early Bird, Syncom and Intelsat.

(7) Photography of the earth's cloud cover from satellites such as Tiros and Nimbus to assist in weather forecasting and the early detection of hurricanes.

(8) The Transit system which provides a global navigation facility for ships and aircraft.

(9) Photography of the planets and the stars from Orbiting Astronomical Observatories (OAO).

(10) Investigation into solar radiation from Orbiting Solar Observatories (OSO).

THE TIDES

The periodic rising and falling of the sea level, commonly known as the *tide*, is primarily caused by the gravitational attraction of the moon and sun acting upon the waters of the earth. The methods employed in predicting the times of the tide at a particular place are extremely complicated and depend to a large extent on the observation of tidal behaviour at the place over a period of many years. In this text the subject of tides will be dealt with quite simply.

Let us consider the tide raising effect of the moon on an earth which is assumed to be uniformly covered with water. It will be remembered that the force of gravitation between two bodies varies directly as the product of their masses and inversely as the square of the distance between their centres. Now, the attractive force of the moon on the water of the hemisphere facing the moon is greater than the attractive force at the earth's centre. Therefore, the water on this hemisphere tends to be drawn towards the line

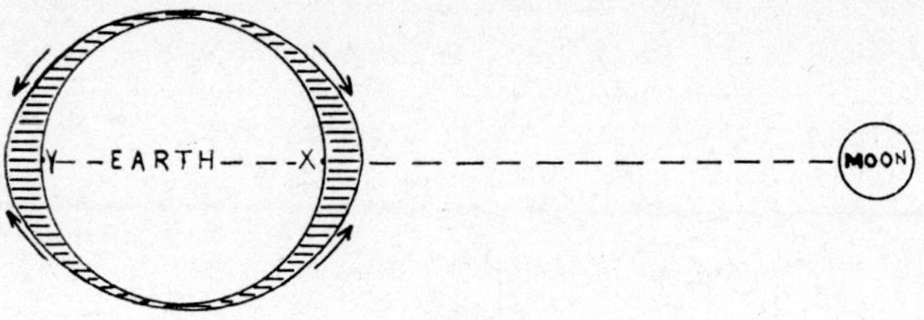

Fig. 5.6
The Moon is the Main Cause of the Tide.

joining the centres of the earth and moon to form a tidal bulge at X (fig. 5.6). The attractive force on the earth's centre is greater than that on the water of the hemisphere turned away from the moon, again causing the water to pile up along the line of centres to give another tidal bulge, this time at Y. Thus, the water on the earth takes on an ellipsoidal shape, being deepest along the earth-moon axis and shallowest in a perpendicular direction. As the earth rotates, these tidal bulges are drawn over the surface water to give any place high tides about 12 hours apart, with a low tide between them.

The sun exerts a tide raising force too, but because it is at such a great distance from earth its effect on the tides is only about half that of the moon. Even so, quite significant differences in tidal levels can occur due to the changing relative positions of these bodies. When the moon is new and full the tidal forces of the moon and sun work together to produce higher high tides and lower low tides, called *spring tides*. At times of quarter moon these forces are out of step with each other, resulting in a small rise and fall in water levels, or *neap tides*.

It was mentioned earlier that actual tidal prediction is an extremely difficult operation. The foregoing explanation takes no account of the effects of the large land masses, variation in depth of the oceans, shallow water near coasts and the earth's rotation. There is also an added complication due to the fact that the sun and moon do not move in the plane of the equator, the declination of the sun ranging from $23\frac{1}{2}°$ N to $23\frac{1}{2}°$ S and that of the moon from $28\frac{1}{2}°$ N to $28\frac{1}{2}°$ S. In spite of these complexities fairly accurate tables of the predicted times and heights of the tides at ports all over the world are published by the Admiralty, mainly for the benefit of shipping.

ECLIPSES OF THE SUN AND MOON

In discussing the apparent motion of the moon on the celestial sphere we saw that its path crosses the sun's apparent path (ecliptic) twice during the lunar month at points called the nodes. This makes it possible for the earth, moon and sun to become in line. When this happens then, since both the sun and the moon have angular diameters of about $\frac{1}{2}°$, either the sun or the moon becomes partially or totally eclipsed as seen from the earth.

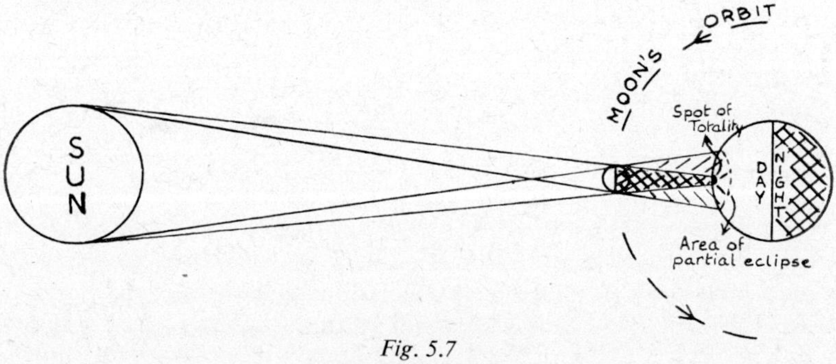

Fig. 5.7
Eclipse of the Sun.

An eclipse of the sun occurs when the new moon passes between the sun and the earth so that the shadow cast by the moon falls upon the earth (fig. 5.7). The dark part of the shadow from which all sunlight is excluded is called the *umbra*, the part from which light is only partially excluded being called the *penumbra*.

The maximum possible diameter of the umbra on earth is about 160 miles (257 km), and if an observer is within the umbra he experiences a *total eclipse* of the sun. The eclipse begins as the moon passes in front of the sun, causing a dark notch to appear in the sun's disc. As the eclipse proceeds, the notch gradually increases in size until the sun is completely hidden by the dark moon. Totality lasts for a couple of minutes, during which time darkness descends, quitness reigns and bright stars become visible. The normally invisible outer atmosphere of the sun, the *corona*, bursts into view as a pearly sheen, presenting a most glorious sight. As the moon moves away the notch in the sun's disc becomes smaller and smaller until the eclipse is over. Due to the movement of the earth and the moon the umbra moves rapidly over the earth's surface, giving rise to a *path of totality*. Astronomers take a great interest in total eclipses and, as they do not occur very often at the same place, travel to remote parts of the world to observe them. The last one to occur in Great Britain was in 1927; the next one will be in Cornwall, near Lands End in 1999. Since the distance of the moon from the earth varies considerably, the umbra may fail to reach the earth so that instead of a total eclipse an *annular eclipse* occurs. The moon then appears smaller than the sun and does not completely obscure it, leaving a bright ring, or annulus, of the sun uneclipsed. To observers in the larger partially shaded area of the penumbra the moon appears to take just a notch out of the sun and they see only a *partial eclipse*. Eclipses of the sun do not occur at every new moon because the shadow may pass above or below the earth.

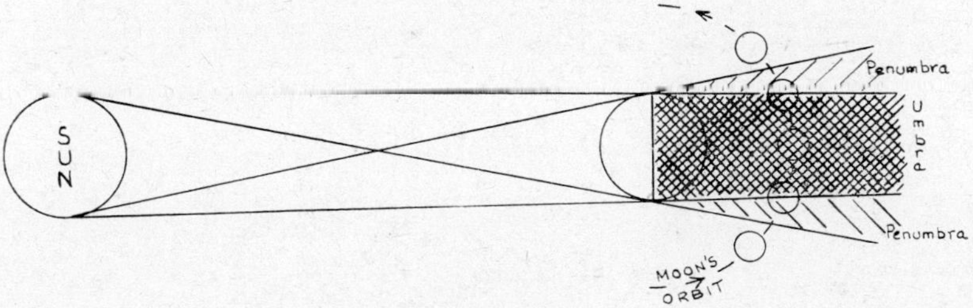

Fig. 5.8
Eclipse of the Moon.

An eclipse of the moon occurs when the moon, at full phase, passes through the earth's shadow (fig. 5.8). The shadow cast by the earth is very much larger than that of the moon, making it possible for the moon to become totally immersed in it for up to

$1\frac{1}{2}$ hours, the eclipse being visible from all parts on the night side of the earth. The moon is darkened slightly as it passes through the penumbra, and as it enters the umbra a dark notch, gradually increasing in size, appears in the moon's disc. At totality it does not disappear completely because light passing through the earth's atmosphere is bent, or refracted, on to the moon, partially illuminating it and giving it a dull coppery glow. If the moon does not become completely immersed in the umbra the eclipse is partial. The moon can miss the umbra and pass only through the penumbra, in which case it simply appears dimmer than usual. More frequently the moon misses the shadow completely so eclipses do not occur at every full moon.

REVIEW QUESTIONS

1. When the moon appears gibbous to us, what is the phase of the earth as seen from the moon?

2. Does every part of the moon experience both daylight and darkness? Is there any twilight on the moon?

3. Does the earth keep the same face turned towards the moon? How much of the earth's surface could be mapped by a lunar observer?

4. Does the earth traverse the lunar sky from one horizon to the other? Do stars rise and set?

5. A lunarnaut gathers 50 lb (22·7 kg) of rock on the moon. How much will this rock weigh on earth?

6. What are the phases of the moon at spring tides and neap tides?

7. Why do high tides recur at any place at intervals of about 12 h 25 m? Can you name any place where unusual tidal effects occur?

8. Is it possible for the earth to totally eclipse the sun as observed from the moon?

9. A spacecraft takes 65 hours to travel the mean distance from the earth to the moon. What is its average speed?

5. The Moon: Last Quarter
Pic du Midi photograph
(*courtesy* PROF. Z. KOPAL).
PLATE III

6. The Moon: First Quarter
Pic du Midi photograph
(*courtesy* PROF. Z. KOPAL).
PLATE IV

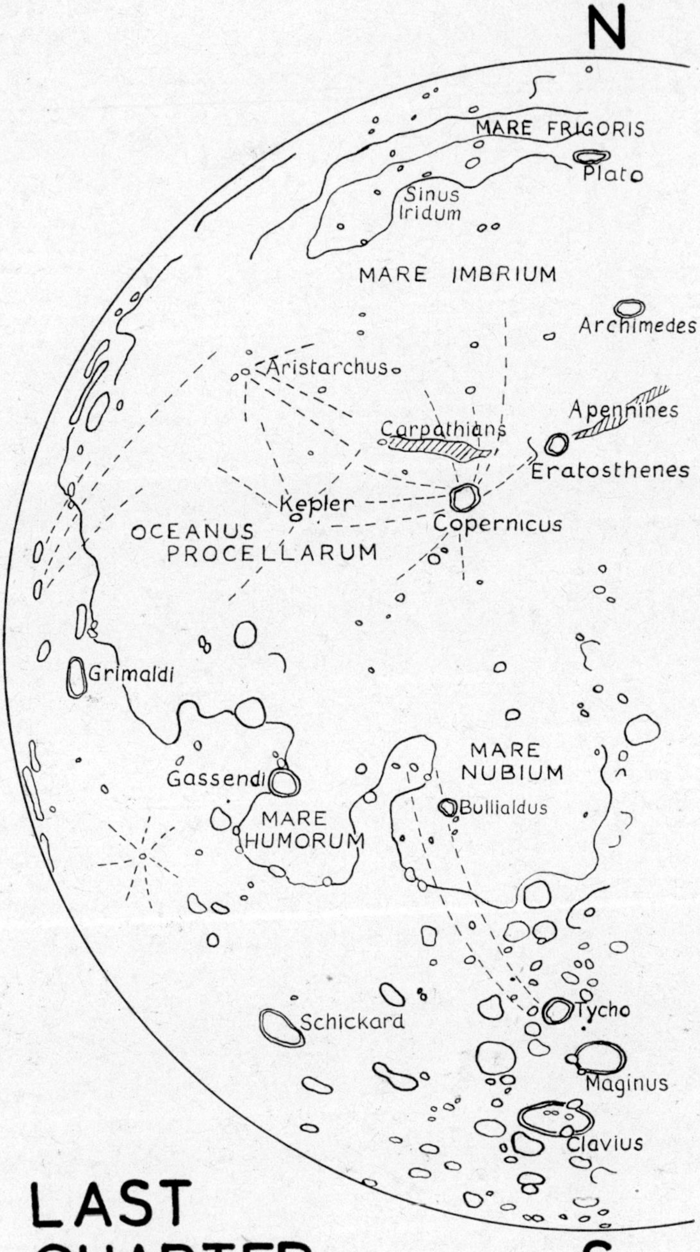

LAST
QUARTER

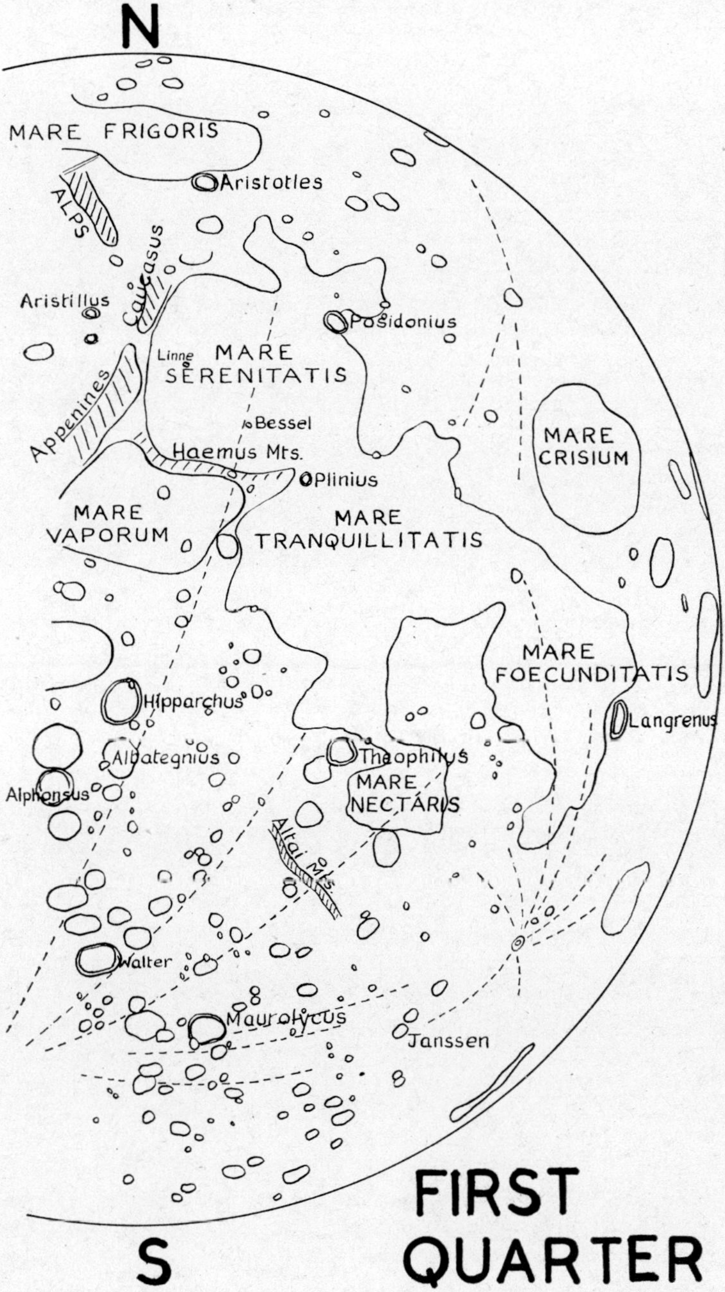
N
MARE FRIGORIS
Aristotles
ALPS
Caucasus
Aristillus
Posidonius
Linne
MARE
SERENITATIS
Appenines
Bessel
Haemus Mts.
Plinius
MARE
CRISIUM
MARE
VAPORUM
MARE
TRANQUILLITATIS
MARE
FOECUNDITATIS
Hipparchus
Langrenus
Albategnius
Theophilus
Alphonsus
MARE
NECTARIS
Altai Mts.
Walter
Maurolycus
Janssen
S
FIRST
QUARTER

PROJECTS AND RESEARCH

1. NAKED-EYE OBSERVATIONS OF THE MOON

On separate sheets of drawing paper prepare a set of faintly drawn circles of 6 in (15 cm) diameter, one for each observation. Starting as soon as possible after new phase, observe the moon on successive nights throughout the lunar month. On each night clearly mark in the position of the terminator and the arc of the circle representing the limb of the moon, after which the remainder of the faint circle can be erased. The rough outlines of the maria may be inserted faintly from the key diagrams to get some uniformity of scale although, in order that the observations be genuine, it is essential to draw only what can be seen. The dark and bright areas can then be lightly shaded with black and yellow pencils respectively. Clearly label the features observed and include the following information with each observation:

(a) Date and GMT of observation.

(b) Age of the moon (the number of days having elapsed since new moon).

(c) Star groups (constellation names) in the immediate vicinity of the moon.

(d) Approximate altitude and azimuth of the moon.

(e) Earthshine when present.

Give a detailed explanation of the causes of the phases of moon and an account of the nature of the dark and bright areas, to be followed by the practical observations arranged in correct sequence. The sheets can then be bound together and suitably labelled according to personal choice.

2. THE MOON'S APPARENT PATH

The object is to prepare in advance a chart showing the apparent path of the moon against the background of the stars and then compare it with visual observations made later. By following the instructions given in Project 4, Chapter 1, a grid system consisting of meridians of Right Ascension and parallels of declination can be made. Having decided in which month it is proposed to carry out observations, extract the values of Right Ascension and declination of the moon from the current edition of Whitaker's Almanack. When these co-ordinates have been plotted on the chart they can be joined up to give a smooth, sinuous curve which represents the moon's apparent path on the celestial sphere for the month. Insert the phases of the moon on the path, together with the dates on which they occur. Also indicate on the chart the positions of some of the bright stars given in the Star Table (Chapter 2). When the time comes around, observe the moon on as many occasions as possible, taking careful note of the stars in the vicinity of the moon, inserting their relative positions on the chart. The ecliptic should also be drawn in as this will verify the fact that the moon's path is inclined at a small angle to it.

3. REGRESSION OF THE MOON'S NODES

A further plotting exercise which may be done either in conjunction with the project described above or independently on a separate chart, will show how the nodes of the moon's apparent path gradually slide in a westerly direction along the ecliptic. Since this regression averages less than 2° a month, it is best to plot the moon's RA and declination for two months separated by as long an interval as possible, say January and December of the same year. When this has been done it will be seen that both the ascending and descending nodes have slewed significantly to the west.

4. PHOTOGRAPHY

The technique of obtaining good photographs of the moon and other astronomical bodies is one which is acquired after long experience and many disappointments. The equipment required can be expensive too. In this book we are concerned with first projects in astronomy, so it is assumed that the reader has at his disposal only a very small telescope or binoculars, and a simple camera. But, even with such modest equipment, reasonably good results can be obtained by photographing the eyepiece image of the moon as described below. Set up the telescope on a rigid stand, or tripod, and direct it towards the moon. Look through the eyepiece and focus the image of the moon until it is at its sharpest. If the camera has a range scale, set it to infinity and then place the camera lens up against the eyepiece lens, with the camera square on to the telescope. If the exposure time cannot be regulated one has to take pot luck, but if it can be, the exposure times should vary between $\frac{1}{30}$ second for a crescent moon and the fastest possible for a full moon. However, the success, or failure, of this exercise may encourage the reader to investigate the more advanced techniques described in some of the books recommended for further reading. See Plate II.

If a telescope is not available there are still a few interesting experiments worth trying. For example, most people have noticed how the full moon appears much larger when it is close to the horizon than it does when riding high in the sky. This is a well known optical illusion, for which there is no adequate explanation. However, the fact that it is an optical illusion can be demonstrated by means of photography. All that needs to be done is to direct the camera at the moon just after it has risen, such that the moon is situated in the bottom left hand corner of the field of view—the diurnal motion of the moon will then carry it towards the top right of the field. Keep the camera perfectly steady and, without winding on the film, make successive exposures of about three seconds in length, at intervals of five minutes. Since excessive use of the shutter may vibrate the camera, it is a good idea to keep it open, with a black card held up against the lens. The card can be removed periodically in order to make the necessary exposures. The result will be a succession of "moons" on the same photograph. After the film has been processed, careful measurement will reveal that the moon's diameter is the same in each case, showing that the moon near the horizon is no larger than when it is higher up. Alternatively, the shutter could be kept open for an hour or so, allowing the moon's image to trail over the film.

Sometimes the moon is seen in the vicinity of a bright planet. On such an occasion direct photographs may be taken, either brief exposures of a few seconds, or longer exposures of several minutes in which case short trails will be obtained.

Should the moon be seen with the silhouette of some interesting terrestrial feature in the foreground, it is a good idea to make a long exposure with the fixed camera so that the moon's trail runs across the photograph. The same sort of thing may also be done with bright star groups.

5. TELESCOPIC OBSERVATIONS

Prepare an outline map of the major features of the lunar surface on a large white card; use the telescope to identify these features. Here, the reader is reminded that an astronomical telescope gives a view which is inverted and also reversed, left to right. Once you have recognised the major features, concentrate on the regions between them and gradually fill in the smaller detail on the card until you have built up a complete map of the moon. Make comprehensive notes on what you see—variations in the size of craters; concentrations of craters; isolated craters; craters possessing ray systems (best observed at full moon); overlapping craters; walled craters; irregularities in the maria; shadows, etc. The amount of detail visible will, of course, depend on the size of telescope being used. Nevertheless, a surprising amount can be seen with a telescope of 1 or 2 inches aperture, or a pair of binoculars.

6. THE EARTH-MOON SYSTEM

At the beginning of this chapter the Earth-Moon system was said to have the character of a double planet. Write an essay concerning these two astronomical bodies, to include:

(a) A comparison of their surface features and general surface conditions.

(b) An account of the effects of (i) the presence of an atmosphere on earth, (ii) the lack of an atmosphere on the moon.

(c) A description of the naked-eye appearance, the diurnal motion, and the movement amongst the stars of each body as seen from the other.

(d) An explanation of phenomena arising from the changing positions of the sun, moon and earth relative to each other—phases, eclipses and tides.

7. THE MOON AND THE TIDES

It is not often possible to link the times of the tides with the moon. However, a closer relation between tidal heights and the moon's phases can usually be detected if the heights of high and

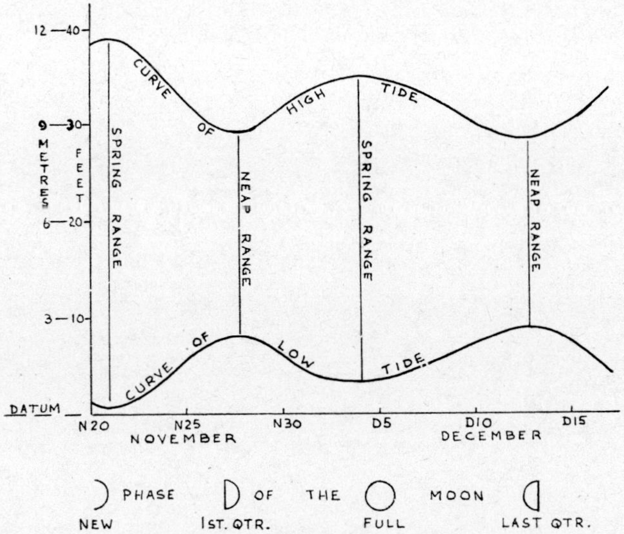

Fig. 5.9
Tides Related to the Moon's Phases.

low tides on each day of the lunar month are plotted on graph paper. A tidal diagram for the Bristol Channel port of Barry is shown in fig. 5.9, on which it can be seen that highest high waters and lowest low waters (spring tides) happen twice a month, on dates corresponding to the occurrence of new moon and full moon, whilst smaller tidal ranges (neap tides) occur between the spring tides at about the times of quarter moon. By similarly plotting curves for several months of the year it may be possible to establish a general correlation between the local tides and the phases of the moon. Tidal information can be obtained from Whitaker's Almanack or, if the reader lives near a port, the local Harbour Master.

8. OBSERVING ECLIPSES

Eclipses of the sun and moon occur fairly often and an opportunity to observe either of these phenomena should not be missed. The dates of their occurrence is given in almanacs and most diaries, although advance publicity is usually given through the media of newspapers and television. No special precautions are necessary when observing lunar eclipses but with regard to solar eclipses the warning concerning the extreme danger of looking at the sun directly is repeated. The best way of observing the sun where a telescope is not being used is through a piece of smoked glass or photographic film which has been exposed and developed. If a telescope is being used, the projection method described in Project 2, Chapter 7, is the only safe way.

9. LIGHT EXPERIMENTS

Make use of the knowledge gained in this chapter to conduct simple experiments in a darkened room to demonstrate the phases of the moon and eclipses of the sun and moon. A torchlight can be used to represent the sun; a tennis ball the earth; and a golf ball the moon. A certain amount of trial and error will be necessary before the correct relative positions of the "sun", "moon" and "earth" are found, particularly in the simulation of eclipses. But, so far as the phases of the moon are concerned it will suffice to move the sun around the moon so that a varying amount of the illumination of the moon can be seen from a given point in the room.

10. OBSERVING OCCULTATIONS

We have seen that although the moon's diurnal motion is from east to west across the sky, it moves from west to east among the stars. This means that the moon often passes in front of stars lying in its path and so blocks them out. Such an event is called an *occultation.* Before full moon, the disappearance of a star takes place at the dark limb and reappearance at the bright limb; the reverse is the case after full moon. Observation of occultations of faint stars with the naked eye are not a practical proposition because they become drowned in the strong light of the moon before they are occulted. Naturally, binoculars or a small telescope are a great asset. Dark limb occultations are most favourable, particularly when the star is bright and the moon is in a crescent phase. Several first-magnitude stars regularly occulted are Aldebaran, Regulus, Pollux, Spica and Antares. The most spectacular feature of a dark limb occultation is the suddenness with which the star disappears. It actually does occur in a split second and is evidence of the moon's lack of atmosphere; otherwise dimming of the star would take place first. Forthcoming occultations are listed in astronomical publications such as the *British Astronomical Association's Handbook* and *Sky and Telescope* monthly magazine. Occultations of planets by the moon, and stars by planets, are possible but very rare.

11. TRACKING ARTIFICIAL SATELLITES

Quite a number of satellites are visible to the naked eye. Whilst the majority are fainter than 3rd magnitude, a few are brighter than this, and Pageos 1 is a rival to any first-magnitude star. In appearance a satellite is starlike, but in practice they are distinguished from stars by their fairly rapid movement across the sky. Since a satellite shines by reflected light from the sun, it means that the most favourable times for observation occur when the satellite is bathed in sunlight, while to the observer the sky is dark. Thus it will be seen to traverse a greater arc of the sky in the hours immediately after sunset and before sunrise. Reference to fig. 5.5 will show that this is more likely to occur when the ground track is near to the observer. The crossing times of the brightest satellites are given in some national newspapers. Predictions are also available from the Satellite Section of the British Astronomical Association, and the Radio and Space Research Station at Slough, though they normally work on an information-observation basis. Here the newcomer to astronomy has an opportunity to do something of scientific value. The usual method of observation is to fix the position of the satellite against the background of the stars by means of a stop-watch, which is started at the instant the satellite is in line with two stars and then checked against the Post

Office telephone speaking clock. For example, an observation might be described as follows: satellite X passed between stars A and B, at $\frac{1}{5}$th their distance apart from A, at 21 h 06 m 32·5 s Universal Time (UT).

12. LUNAR EXPLORATION

The 1960s will probably be remembered as one of the most exciting decades in the history of mankind. Following the flight around the moon of Apollo 8, astronauts Stafford, Young and Cernan took their craft (Apollo 10) to within 10 miles (16 km) of the moon. Then on July 16, 1969, Armstrong, Aldrin and Collins set off on man's first expedition to another astronomical body. While Collins circled the moon in the mother-craft, Armstrong and Aldrin descended in the lunar excursion module (LEM) to become the first men to set foot upon the moon. They brought moon rock back to earth, leaving behind several experiments and the memorable plaque which read:

"Here men from the planet Earth first
set foot upon the Moon
July 1969 A.D.
We came in peace for all mankind"

On November 14, 1969, Apollo 12 with astronauts Conrad, Bean and Gordon aboard mounted the second expedition to the moon. Others will follow.

Making full use of school, college and public libraries, prepare an account of "Man's Conquest of the Moon". Use the list given below as a basis for your research.

ANSWERS TO REVIEW QUESTIONS

1. Crescent.
2. Yes. No.
3. No. All.
4. No. Yes.
5. 300 lb (136·2 kg).
6. New and Full. Quarter.
8. Yes.
9. 3677 mph (5916 km/h).

TABLE 5.1
HIGHLIGHTS OF LUNAR EXPLORATION

SPACECRAFT	LAUNCH DATE	COUNTRY	REMARKS
	1959		
Lunik 2	Sept. 12	USSR	First probe to hit the moon.
Lunik 3	Oct. 4	USSR	Photographed far side of moon.
	1964		
Ranger 7	July 28	USA	Close-up photographs before impact.
	1965		
Ranger 8	Feb. 17	USA	Photographs before impact. Also Ranger 9. (21.3.65).
	1966		
Luna 9	Jan. 31	USSR	First soft landing. Photographs.
Luna 10	Mar. 31	USSR	Achieved lunar orbit.
Surveyor 1	May 30	USA	Soft landing. Photographs.
Lunar Orbiter 1	Aug. 10	USA	Orbited moon. Photographs. Also Lunar Orbiter 2. (6.11.66).
Luna 12	Oct. 22	USSR	Orbited moon. Photographs.
Luna 13	Dec. 21	USSR	Soft landing. Photographs and tests.
	1967		
Lunar Orbiter 3	Feb. 5	USA	Orbited moon. Photographs. Also Lunar Orbiter 5 (1.8.67).
Surveyor 3	Apr. 17	USA	Soft landing. Photographs and tests. Also Surveyor 5 (8.9.67), Surveyor 6 (7.11.67).
Explorer 35	July 19	USA	Tested moon's space environment.
	1968		
Surveyor 7	Jan. 7	USA	Soft landing. Photographs and tests.
Zond 5	Sept. 21	USSR	Returned experiment package to earth after orbiting moon.
Apollo 8	Dec. 21	USA	Men orbit the moon.
	1969		
Apollo 10	May 18	USA	Men descend almost to the moon's surface.
Apollo 11	July 16	USA	Men land on the moon. Rock samples brought back to Earth.
Apollo 12	Nov. 14	USA	Second manned expedition to the moon. Returned with rock samples and part of Surveyor 3.
	1970		
Apollo 13	Apr. 11	USA	Mission aborted after on-board explosion. Astronauts orbited Moon and returned safely to Earth.
Luna 16	Sept. 12	USSR	Scooped up small quantity of moon dust and returned to Earth (unmanned).
Luna 17	Nov. 10	USSR	Landed small robot car on the moon.

CHAPTER 6

FAMILY OF THE SUN

The sun and the host of smaller bodies revolving around it are known collectively as the solar system. The solar family comprises nine major planets—Mercury, Venus, Earth, Mars, Jupiter, Saturn, Uranus, Neptune and Pluto—with their attendant satellites; a multitude of small rocky bodies called asteroids, or minor planets, circling the sun between the orbits of Mars and Jupiter; comets; and meteor streams.

THE PLANETS IN GENERAL

The orbits followed by the planets are ellipses, though it must be stressed that except in the cases of Mercury and Pluto these ellipses do not depart greatly from circles (fig. 6.1). In Chapter 3 we saw that the earth revolves and rotates from west to east. The other planets not only perform their revolutions and rotations

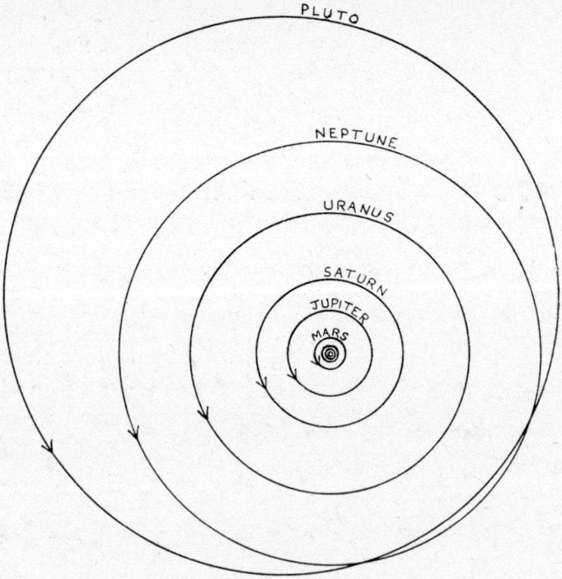

Fig. 6.1
Orbits of the Major Planets.

in the same direction, their orbits too are very nearly in the same plane as that of the earth's. This means that as observed by us they never stray very far from the ecliptic and, with the occasional exception of Venus and Pluto, are always to be found within the limits of the zodiac.

The time taken for a planet to make one complete revolution around the sun is its year, or *sidereal period.* The further away a planet is from the sun the slower it revolves and the longer its year. A year on Mercury is only 88 earth days in length, but on Pluto an imaginary inhabitant having the average life span of an earth-being would barely reach the "ripe old age" of $3\frac{1}{2}$ months, for Pluto's sidereal year is 248 times longer than ours.

The *synodic period* is the interval between two successive overtakings of the earth by a faster moving inner planet, or the interval between two successive overtakings of a slower moving outer planet by the earth. Jupiter and the planets beyond it move slow enough to be overtaken by the earth at intervals of a little over a year. Fast moving Mercury has the shortest synodic period, 116 days; Venus and Mars have the longest, 584 and 780 days respectively.

THE INFERIOR PLANETS

The planets Mercury and Venus, whose orbits are inside that of the earth's, are known as the *inferior planets.* Because they revolve faster, the inferior planets gain on the earth, appearing to us sometimes to lead the sun in the sky and sometimes to lag behind it. When an inferior planet is to the west of the sun it sets before the sun, but since it rises before the sun it can be seen as a "morning star" towards the east. When it is to the east of the sun it remains above the western horizon after sunset as an "evening star". An inferior planet is said to be at *inferior conjunction* when the earth, the planet, and the sun are in the same line with the planet between the earth and the sun, at which time it makes its closest approach to the earth. It is at *superior conjunction* when the earth, the planet, and the sun are again in the same line, this time with the planet on the far side of the sun from the earth. Like the moon the inferior planets pass through a succession of phases, the cause of which is easily understood by referring to fig. 6.2. Consider an inferior planet at superior conjunction: after passing this position it emerges to the east of the sun to become an evening star in gibbous phase. The angle at the earth contained between the directions of the sun and the planet (*elongation*) steadily increases to reach a maximum when the planet exhibits a quarter phase.

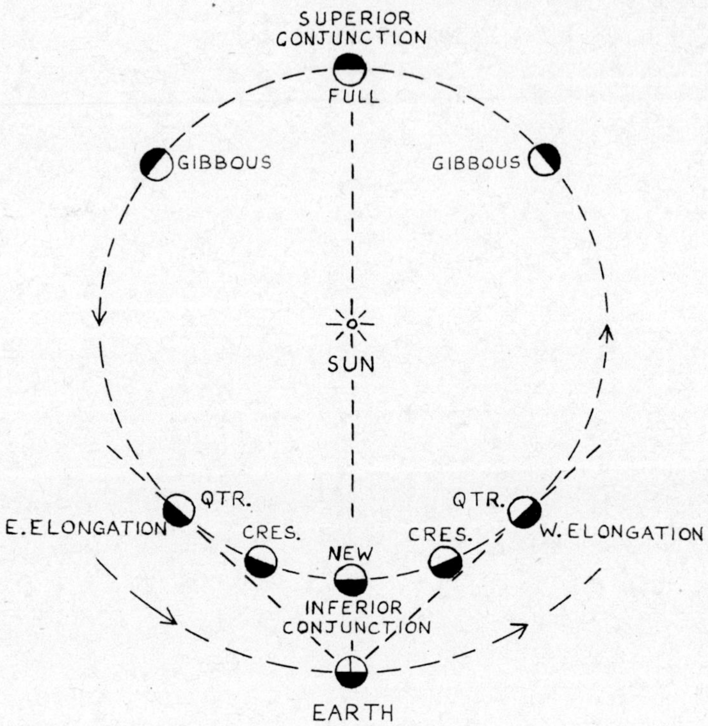

Fig. 6.2
The Changing Aspect of an Inferior Planet.

After maximum eastern elongation the phase becomes a shrinking crescent as the planet appears to move rapidly between the earth and the sun. On passing inferior conjunction the planet moves to the west of the sun and into the morning sky. The phase increases from crescent to quarter at maximum western elongation, thence to gibbous phase until superior conjunction is again reached. Greatest elongation does not exceed 48° for Venus, and 28° for Mercury. The small elongation of Mercury means that it is always close to the sun in the sky, making it a very elusive object to observe. On the other hand, Venus, with its greater elongation, larger size and closer approach to the earth, becomes a dominant feature of the evening or morning skies either side of inferior conjunction.

The inferior planets usually pass north or south of the sun at inferior conjunction, but just occasionally either Mercury or Venus

7. The Largest Optical Telescope in Great Britain
The 98-in (2·54-m) Isaac Newton reflecting telescope, prior to its installation at the Royal Greenwich Observatory.
(*Courtesy* GRUBB PARSONS LTD.).

8. The Surface of Mars
Note the resemblance of the Martian surface to that of the Moon. Photographed by Mariner 6 on July 30, 1969, at a range of 2150 miles (3460 km). The largest crater visible is about 160 miles (256 km) in diameter.
(*Courtesy* JET PROPULSION LABORATORY, NASA).

PLATE V

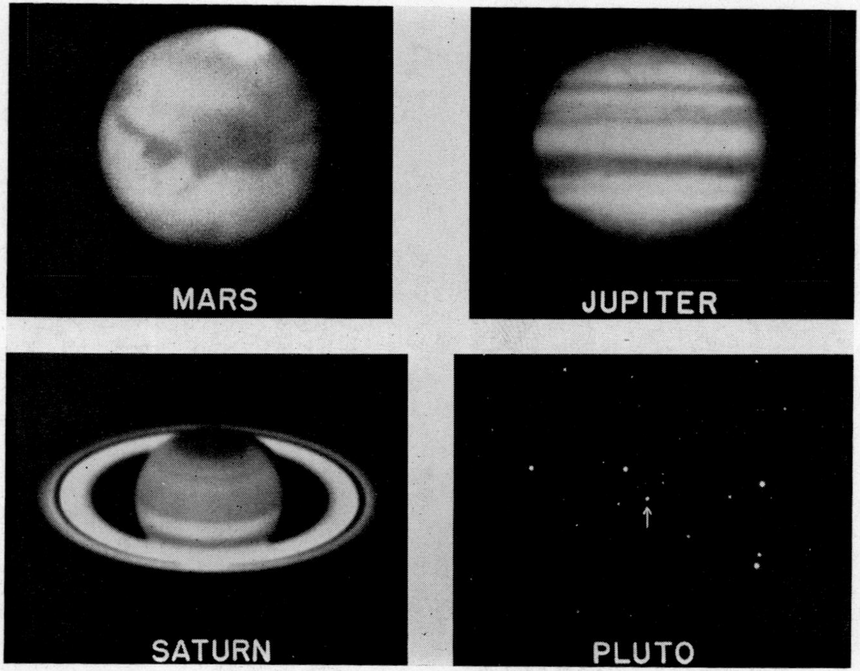

9. Mars, Jupiter, Saturn and Pluto
The first three were photographed with the 100-in (2·54-m) telescope, Pluto with the 200-in (5·08-m).
(*Courtesy* MOUNT WILSON *and* PALOMAR OBSERVATORIES).
PLATE VI

transits directly in front of the sun to appear as a dark dot against the bright solar disc. Transits of Mercury occur about 13 times in every 100 years, the next one being November 9, 1973. Transits of Venus are much less common, the next one occurring in 2004.

THE SUPERIOR PLANETS

The planets whose orbits lie outside that of the earth's—Mars, Jupiter, Saturn, Uranus, Neptune and Pluto—are called the *superior planets.* Since they revolve more slowly than the earth does, the earth gains on the superior planets so that although they have a general eastwards motion with respect to the distant stars, they do in fact have a westerly motion with respect to the sun. A superior planet is at *conjunction* when the earth, the planet, and the sun are in the same line with the planet on the far side of the sun from the earth. At *opposition* they are again in line, but with the earth in between the planet and the sun. Superior planets make their closest approach to the earth at opposition and this is the best time to observe them, especially Mars.

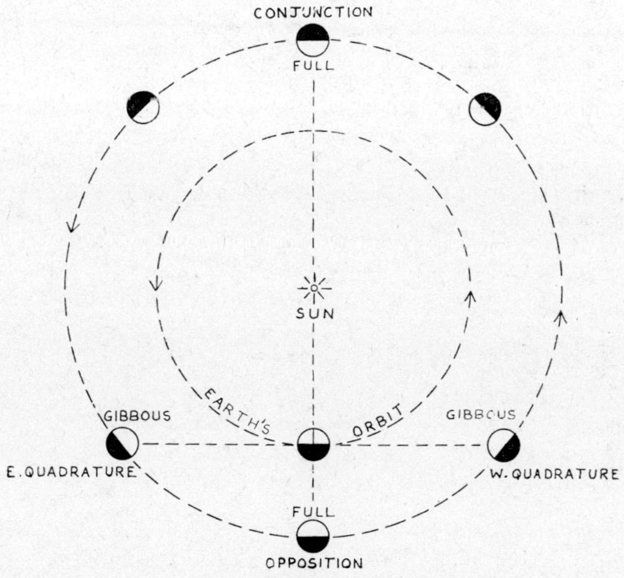

Fig. 6.3
The Changing Aspect of a Superior Planet.

Referring to fig. 6.3, it can be seen that in the case of a superior planet elongation varies between 0° at conjunction to 180° at opposition. At conjunction a superior planet is on the meridian with the sun at noon. After conjunction it emerges to the west of the sun as a morning star, rising earlier than the sun from day to day. At *western quadrature* (elongation 90°) the planet is near the meridian at the time of sunrise. At opposition the planet is opposite the sun and is on the observer's meridian at midnight. After opposition it continues its westerly movement with respect to the sun, setting earlier each night. At *eastern quadrature* it is near the meridian at sunset. Then, as it gradually approaches the next conjunction it becomes lost in the evening twilight.

The superior planets do not go through a complete cycle of phases like the moon; at both opposition and conjunction they exhibit a full phase, and in other positions they do not depart much from full, except for Mars which is distinctly gibbous at quadrature.

APPARENT MOTIONS AMONG THE STARS

The planets revolve around the sun from west to east, all having a general eastwards or *direct motion* with respect to the stars. However, if successive positions of a planet as it moves against the stars were carefully recorded it would be found that at a certain time the planet would appear to turn about and move westwards, or *retrograde*, for a while before resuming its eastwards motion. The points where it "stands still" before retrograding and before resuming normal eastwards motion are called *stationary points*. The retrograde loop performed by the planet Mars during 1967 is shown in fig. 6.4. Retrograde motion completely baffled early

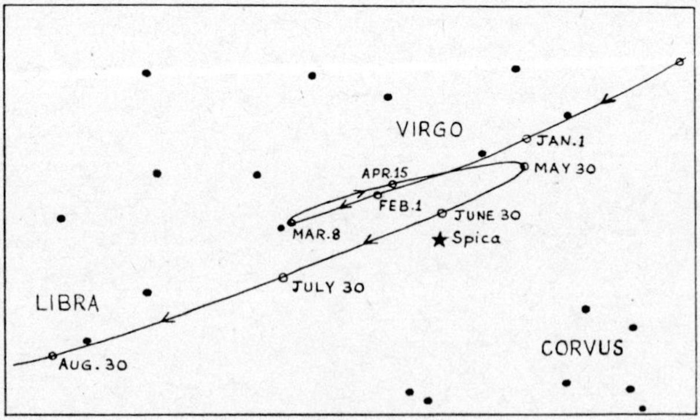

Fig. 6.4
The Path of Mars among the Stars.

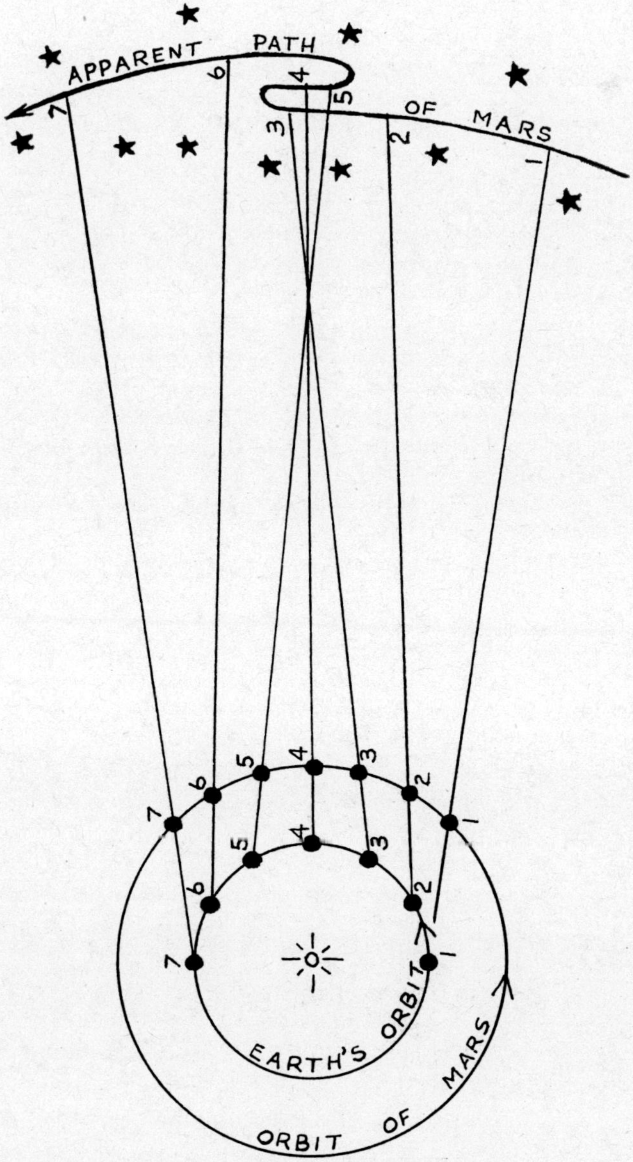

Fig. 6.5
Retrograde Motion Explained.

astronomers, yet the cause of it is simply a matter of the changing perspective of a planet with reference to the more distant stars at a time when the planet is nearest to the earth. Fig. 6.5 illustrates how the retrograde loop of a superior planet, Mars, is caused when the planet is being overtaken by the earth near to the time of opposition. If the outer circle of the figure is taken as the earth's orbit and the inner circle the orbit of an inferior planet, say Venus, the lines of sight projected in the reverse direction will show that an inferior planet retrogrades as it is overtaking the earth near to the time of inferior conjunction.

PROGRESSIVE IDEAS ON PLANETARY MOTIONS

The word planet (meaning wanderer) was originally applied to Mercury, Venus, Mars, Jupiter and Saturn by early astronomers, to whom the movements of these bodies against the background of the stars were well known, though not understood. These early astronomers believed the earth was a stationary object surrounded by a large hollow sphere upon which the stars were fixed. The sphere was assumed to rotate once every day, causing the stars to rise and set. Within the star sphere the Moon, Mercury, Venus, Sun, Mars, Jupiter and Saturn, in order of distance, moved in a general eastwards direction in circles around the central earth. In this *geocentric system*, it was difficult to account for the occasional retrograde motion of the planets, one of the earliest solutions to the problem being found in *Ptolemy's Almagest* (about 150 A.D.).

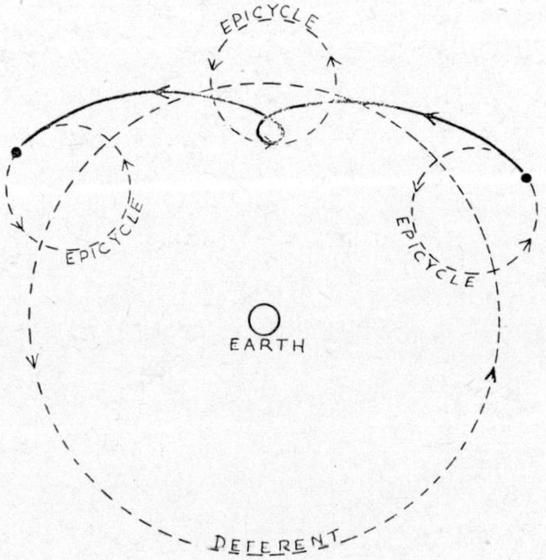

Fig. 6.6
Early Idea of Planetary Motion.

In the Ptolmaic system each planet moved uniformly in a circle (the *epicycle*), whilst the centre of this circle itself moved in a circle (the *deferent*) around the earth. Fig. 6.6 shows in a simple form how these two circular motions combine to give the necessary looped path of a planet against the starry background; although it seemed quite in order to add further epicycles wherever it was required to keep the system in accord with the observed motions of the planets.

The Ptolmaic system went unchallenged for 1400 years until Nicolas Copernicus (1473-1543), a Pole of German descent, sowed the seeds of doubt. In his book, *On the Revolutions of Celestial Bodies*, Copernicus stated that these movements could be more easily explained by assuming the central sun, around which the planets, including the earth, revolved (*heliocentric system*). His suggestion that the earth rotated on its axis would also account for the rising and setting of the stars and other celestial bodies. He still assumed circular orbits for the planets and clung to the use of epicycles. Nevertheless, in spite of the opposition encountered by the Copernican system at the time, this great break with traditional thought is nowadays generally regarded as the beginning of modern astronomy.

Tycho Brahe (1546-1601) was a Danish Astronomer, who, with the financial aid of King Frederick of Denmark, established an observatory on the island of Hveen. After the death of the king, Tycho fell into disfavour with the ruling nobility and the last years of his life were spent at Prague. Although he gained great fame as an astrologer, he was an extremely accurate observer and his excellent observational records were to be of great use to later astronomers, particularly Johannes Kepler (1571-1630), a German, who came to work with Tycho. When Tycho died he bequeathed all his observational data to Kepler who used it, especially the records of Mars, as a basis for his famous *laws of planetary motion*. Careful study of Tycho's observations and further experimenting led Kepler to dispense entirely with the use of epicycles and to the conclusion that the planets moved around the sun not in circles, but ellipses. Kepler's Laws are as follows:

1. Each planet revolves around the sun in an elliptical orbit with the sun at one focus of the ellipse.
2. Each planet revolves around the sun so that the line joining it to the sun sweeps out equal areas in equal times.
3. The square of the time taken by each planet to complete its orbit is proportional to the cube of its mean distance from the sun.

Thus Kepler destroyed completely the older ideas that planetary movements could only be represented by combinations of circular motions.

While Kepler was engaged in his studies, Galileo Galilei (1564-1642), the son of a Florentine nobleman, was finding further evidence in support of the Copernician system. The first astronomical telescope was made and used by Galileo, and with it he discovered four bright satellites of Jupiter, the mountains on the moon, the rings of Saturn, the phases of Venus, and sunspots. The discovery of the satellites revolving around Jupiter squashed the old idea that the moon would be left behind if the earth really revolved around the sun. Also, from the movement of sunspots, Galileo inferred the sun's rotation and by analogy the earth's rotation. As a result of his observations he became a champion of the Copernican system but this, together with his findings on sunspots, led him into deep trouble. Persecuted by the Church, he was forced to give up his academic post; yet even during his old age, when nearly blind, he discovered the librations of the moon.

The solution of the problem of planetary motions expressed by Kepler's Laws, is a purely geometrical one. Galileo did some pioneer work on the motions of bodies in general, but it was the illustrious Sir Isaac Newton (1642-1727) who introduced the concept of forces acting throughout the universe. From his studies of the motion of the moon he inferred the existence of the force of gravity, his great work on the motions of bodies (dynamics) being set out in *The Principia,* published in 1687. By means of these principles he was able to prove Kepler's Laws and to reduce Kepler's geometrical description of the planetary system to a single physical law, which states: "the force between any two bodies in the universe is directly proportional to the product of their masses and inversely proportional to the square of the distance between their centres." With the passing of time, the universality of his law has been challenged in some respects, particularly where bodies travelling with great velocities are concerned. Even so, modern celestial mechanics is still based to a large extent on Newton's work.

THE PLANETS

The planets fall into two distinct groups—the terrestrial planets, and the giant planets. Those of the terrestrial group, which includes Mercury, Venus, Earth and Mars, are nearest to the sun. They are characterised by their small sizes and great diversity in surface conditions; each possesses a solid core, and all but Mercury have significant atmospheres. The giant planets are Jupiter,

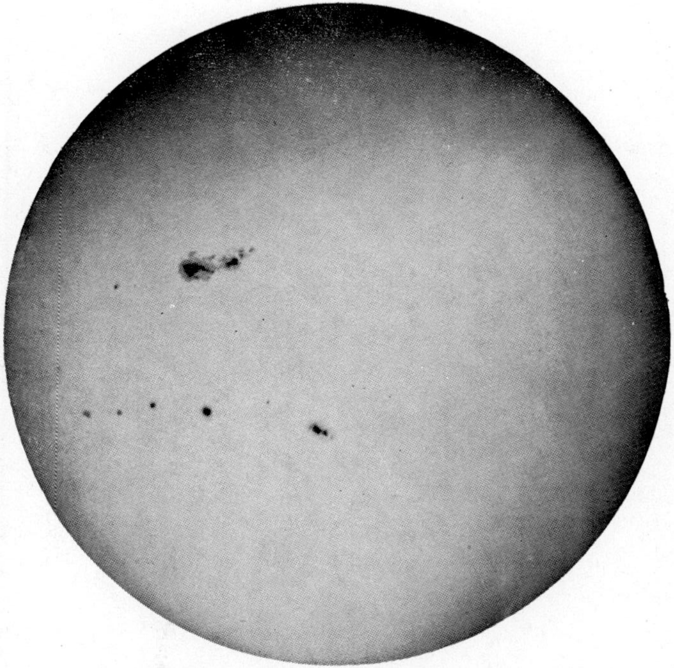

10. Comet Arend-Roland (1956h)

11. The Sun at Sunspot Maximum

(Left) this remarkable photograph of Comet Arend-Roland, taken on April 27, 1957, shows an apparent sunward spike. (Right) the Sun on July 11, 1969, sunspot maximum year. Both photographs were taken by FRANK J. ACFIELD *from his observatory at Forest Hall, Newcastle upon Tyne.*

PLATE VII

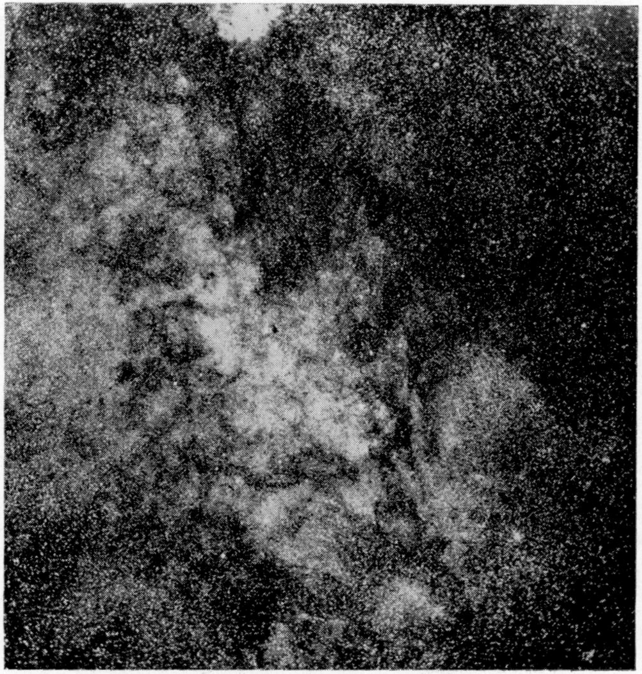

12. A Section of the Milky Way — *13. A Bright Galactic Nebula*

(*Left*) *the Great Star Cloud in Sagittarius, photographed by* E. E. BARNARD (*courtesy* ROYAL ASTRONOMICAL SOCIETY).
(*Right*) *NGC1976*(*M42*) *The Great Nebula in Orion, photographed with the 100-in* (*2·54-m*) *Hooker telescope*
(*courtesy* MOUNT WILSON *and* PALOMAR OBSERVATORIES).

PLATE VIII

Saturn, Uranus and Neptune. They are noted for their large sizes, rapid rotations, and deep atmospheres consisting of hydrogen, methane and sometimes ammonia. So little is known about the outermost planet, Pluto, that it is difficult to associate it with either group.

TABLE 6.1
PLANETARY DATA

PLANET	MEAN DISTANCE FROM SUN (A.U).	(million miles)	(million km)	ORBITAL PERIOD	INCL. TO ECLIPTIC (degrees)	ROTATION PERIOD	EQUATORIAL DIAMETER (miles)	(km)
MERCURY	0.39	36	58	88 d	7.0	59 d	3010	4840
VENUS	0.72	67	108	225 d	3.4	244 d	7650	12300
EARTH	1.00	93	150	$365\frac{1}{4}$d	—	23 h 56 m	7926	12756
MARS	1.52	142	228	687 d	1.8	24 h 37 m	4220	6790
ASTEROIDS (Ceres)	2.77	257	414	4.6 y	10.6	—	430	700
JUPITER	5.20	484	778	11.9 y	1.3	9 h 50 m	88 700	142800
SATURN	9.54	887	1427	29.5 y	2.5	10 h 14 m	74 200	119500
URANUS	19.18	1783	2870	84.0 y	0.8	10 h 49 m	29 300	47100
NEPTUNE	30.05	2794	4499	164.8 y	1.8	15 h 48 m	27 900	44800
PLUTO	39.52	3670	5900	247.7 y	17.2	6.4 d	3670	5900

MERCURY

Mercury, equatorial diameter 3010 miles (4840 km), is the smallest of the major planets. It revolves around the sun in the short period of 88 days at a mean distance of 36 million miles (58 million km), but due to the high eccentricity of its orbit Mercury's distance from the sun ranges from 29 million miles (47 million km) at perihelion to 43 million miles (69 million km) at aphelion. Observation of the planet is very difficult, for even when maximum elongation occurs at aphelion the angular distance of the planet from the sun does not exceed 28°. The game of celestial hide and seek it plays, makes the name Mercury, "the winged messenger", a very appropriate one for this planet. At its best Mercury attains a magnitude of —1·1, second only to the brightest star Sirius. Large telescopes reveal some surface markings, but these are generally faint and ill-defined. Consequently, astronomers have found great difficulty in determining its rotational period. It was previously thought to be the same as the revolution period so that Mercury always kept the same face turned towards the sun. However, in 1965, American astronomers using radar techniques established that Mercury rotates on its axis once every 59 days. Because of its nearness to the sun, Mercury is very hot, the temperature of its sun-scorched surface rising to about 400°C; high enough to melt tin and lead. In many ways Mercury resembles our own moon; its low gravity and high temperature make it unlikely that it possesses any water or atmosphere; its reflecting power of sunlight is also about the same as that of the moon, indicating the likelihood of a rough, irregular surface. Mercury has no satellites.

VENUS

Venus is the second major planet outwards from the sun, around which it travels in a near circular orbit at a mean distance of 67 million miles (108 million km), taking 225 days to complete one revolution. It is familiar to us as a magnificent morning or evening star, and at times of greatest brilliancy it outshines all other celestial bodies apart from the sun and moon. It is brightest when in the crescent phase about five weeks before and after inferior conjunction, and can attain a magnitude of —4·4. With the exception of the moon, Venus is our nearest astronomical neighbour, approaching within 26 million miles (42 million km) of the earth at inferior conjunction. The phases are the planet's most conspicuous

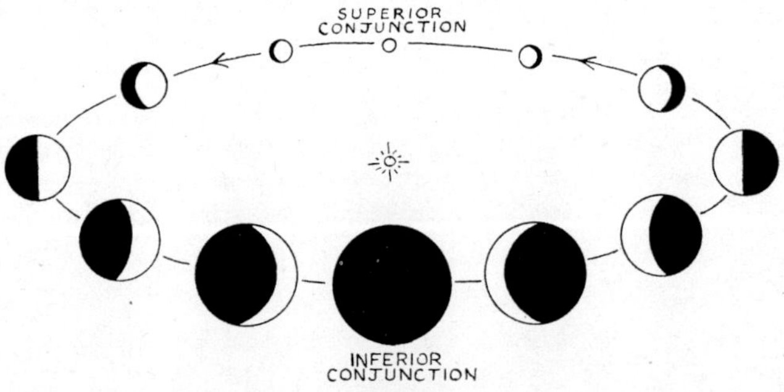

Fig. 6.7
The Changing Phases of Venus.

feature when seen through the telescope, and its apparent diameter when near inferior conjunction is about eight times that when in the gibbous phase near superior conjunction (fig. 6.7). Venus is sometimes known as the earth's twin: their equatorial diameters are 7650 miles (12 300 km) and 7926 miles (12 756 km) respectively. They are also similar in mass and surface gravity but their atmospheres are totally different. The Venusian atmosphere consists of dense clouds, estimated to be 55 miles (90 km) in depth, which completely obscure the surface of the planet. Oxygen and water vapour have been detected, though not in very appreciable quantities. It is thought that nitrogen forms a large part of the atmosphere, but in the light of recent investigations there appears to be an abundance of carbon dioxide. Surface temperatures, estimated previously to be around 100°C, are now thought to be very much greater; probings of the planet by American and Russian spacecraft suggest it may be as high as 700°C. The surface is probably a dry, dusty desert, completely devoid of water, but because of the opaqueness of its atmosphere Venus is still very much a mysterious planet. It is the absence of any observeable permanent features that makes the rotation period difficult to determine. After years of confusion, and estimates ranging from as little as 12 hours to as much as 225 days, there does seem to be some evidence to support a rotation period of 244 days. If this is true, it means that a day on Venus is longer than the year. There is still much to be learned about the veiled planet, and it almost certainly holds many surprises in store for us.

MARS

Mars, the next planet beyond the earth, orbits the sun at a mean distance of 142 million miles (228 million km), taking 687 days to complete one revolution. Its equatorial diameter, 4220 miles (6790 km), is just over half that of the earth's, and a Martian day, 24 h 37 m, is only slightly longer than our own. Being a superior planet Mars is most favourably placed for observation at opposition, when it is visible throughout the night, successive oppositions occurring at intervals of approximately 780 days. However, because its orbit is very eccentric, its distance from the earth at opposition ranges from 62 million miles (100 million km) at Mars' aphelion to 35 million miles (56 million km) at perihelion. Favourable oppositions occur every 15 or 17 years, and always in August or September (fig. 6.8). Whilst its phase does not depart much from full, its magnitude can vary between —2·8 and + 2·0.

Mars has always held a fascination for astronomers and laymen alike, and more than any other planet has been the subject of controversy for many years. Even moderate telescopes can reveal quite distinct surface features: white polar caps, which are very conspicuous during the winter of a hemisphere, shrink or disappear

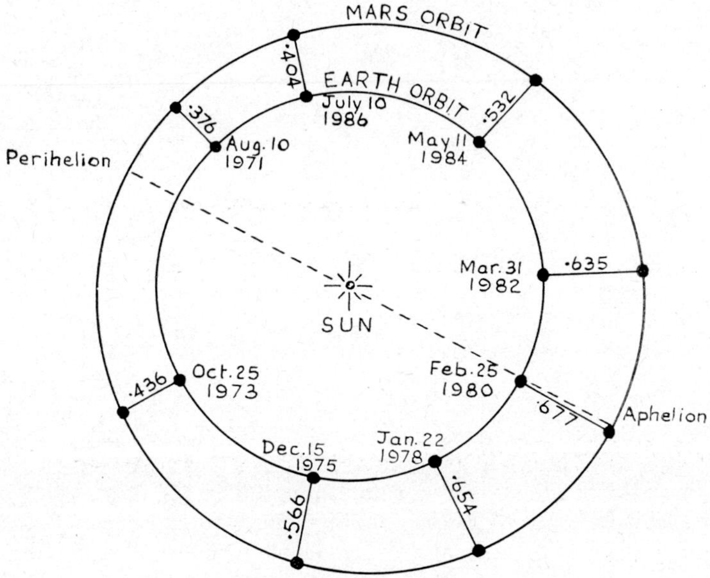

Fig. 6.8
Mars at Successive Oppositions.

with the coming of summer. About 60 per cent of the surface has a reddish hue—probably desert expanses or oxidised rocks—hence the name "The Red Planet". Other darker areas, originally thought to be water, bear such names as Syrtis Major (Great Bog), Solis Lacus (Lake of the Sun) and Mare Sirenum (Sea of Syrens). The spin axis of Mars is inclined at 25° to the plane of its orbit, nearly the same as that of the earth, so that its seasons are like ours except that they are almost twice as long. With the onset of spring in a hemisphere, the polar cap starts to contract and a darkening extends over the Martian surface from the polar regions towards the equator. Towards the end of summer the dark markings fade and recede back to the pole, where the snow cap again assumes larger proportions. While the southern cap often disappears completely, the northern cap is almost always in evidence. These colour changes have been associated with the growth and decay of vegetation and has opened up a great controversy on whether or not some form of life exists on Mars. In 1877 an Italian astronomer, Schiaparelli, announced the existence of a network of fine lines extending over the Martian surface. Unfortunately, he called them "canali" which was im-

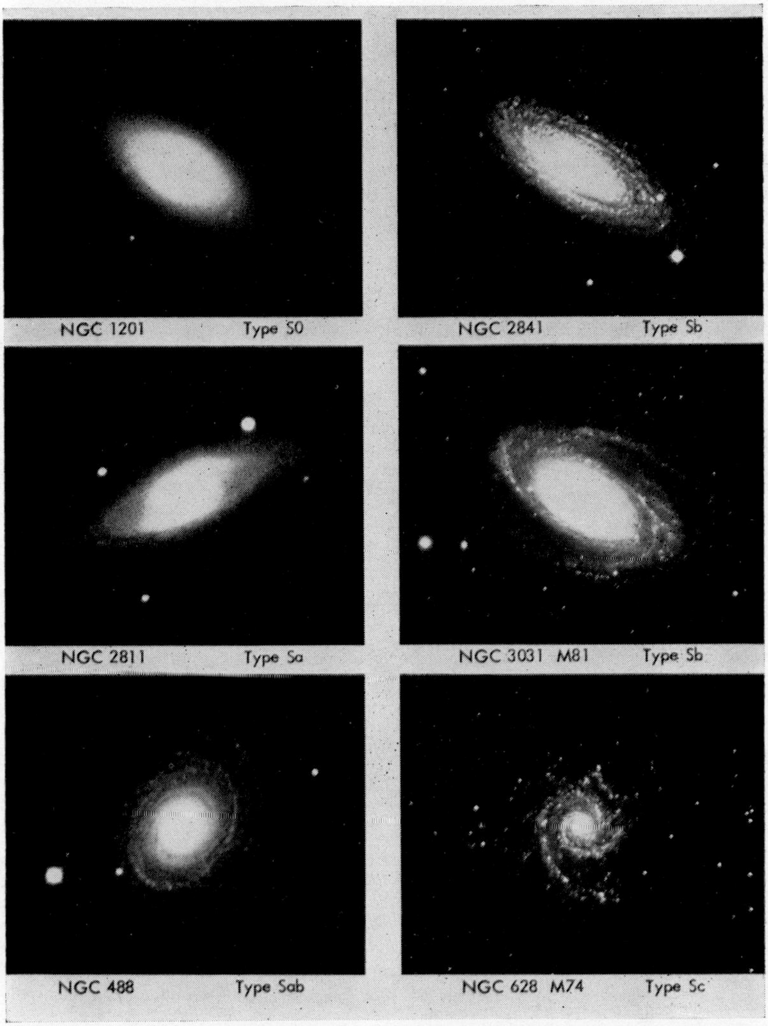

14. Six Normal Spiral Galaxies
(*courtesy* MOUNT WILSON *and* PALOMAR OBSERVATORIES).
PLATE IX

15. A Globular Cluster

16. A Radio Telescope

(*Left*) *NGC6205(M13) the Globular Cluster in Hercules, photographed with the 200-in (5·08-m) Hale telescope*
(*courtesy* MOUNT WILSON *and* PALOMAR OBSERVATORIES).
(*Right*) *the 82-ft (25-m) fully steerable satellite tracker and radio telescope at the Radio and Space Research Station, Chilbolton, Hampshire*
(*courtesy* AEI ELECTRONICS).

PLATE X

mediately interpreted to mean canals. Once the idea was implanted in the minds of other observers, confirmation of the canals' existence was readily forthcoming. Percival Lowell, an American, was so convinced that the canals were artificial waterways constructed by the Martians to irrigate their dry planet, that he established an observatory at Flagstaff, Arizona, specially for the purpose of studying Mars. Although the idea of canals is still bandied about in some quarters, in scientific circles their existence has long been discredited. Close-up photographs from spacecraft (Mariner 4, 1965, and Mariners 6 and 7, 1969) failed to reveal any evidence of canals; rather, they showed many hundreds of craters in the Martian surface, indicating that the planet may not be greatly different from that of the moon.

The atmosphere of Mars is much less dense than ours, and many of the visible clouds are probably dust blown up from the desert regions. It is known to contain carbon dioxide, but the water and oxygen content is thought to be very small; thus the polar caps are probably no more than light coverings of frost. The thin atmosphere can afford little protection from dangerous solar radiation, which implies that surface conditions must be very harsh. The average surface temperature is estimated to be 50°C lower than that existing on earth. At the equator the range of temperature from day to night is in the region of 100°C, so that the only types of vegetation likely to survive such rigorous conditions are primitive mosses or lichens such as are found in the Arctic regions on earth.

Mars possesses two mini-satellites, Phobos and Deimos, 10 miles (16 km) and five miles (8 km) in diameter respectively.

THE ASTEROIDS

The asteroids, otherwise known as *minor planets* or *planetoids*, are small celestial bodies circulating around the sun mainly between the orbits of Mars and Jupiter. The first asteroid, Ceres, was discovered accidentally in 1801 by the Italian astronomer Piazzi, and within the next few years three more, Pallas, Juno and Vesta were found. From 1845 onwards further discoveries of asteroids followed rapidly, particularly after the application of photography to the search in 1891. Since then several thousand have been located, and each year many more are picked up in photographs, often in the course of other investigations. The asteroids range in size from small fragments to 430 miles (700 km) in diameter in the case of Ceres, the largest. They are mostly irregular in shape and Vesta is the only one capable of being seen with the naked eye.

The majority have nearly circular orbits slightly inclined to the ecliptic, with periods of revolution around the sun of between three and a half and six years. A notable exception is Hidalgo, whose orbit is inclined at 43° to the ecliptic and is so eccentric as to extend as far as the orbit of Saturn. Adonis, Apollo and Hermes come within the orbit of Venus, and Icarus passes within the orbit of Mercury. In 1968, Icarus whizzed past the earth at a distance of only 10 times that of the moon. More than a dozen asteroids form a special group known as the Trojans, which move in two clusters in the orbit of Jupiter, one ahead of and the other behind the planet. One theory suggests that the asteroids are fragments of rock resulting from the explosion of a planet during the early life of the solar system. If this theory is correct the planet must have been a very small one because the total mass of the asteroids is estimated at only 1/1000 that of the earth.

JUPITER

Jupiter, the "noble planet", is the largest body in the solar system apart from the sun itself. Its vital statistics are:

equatorial diameter	88 700 miles (142 800 km)
polar diameter	82 800 miles (133 500 km)

It orbits the sun at a mean distance of 484 million miles (780 million km), taking almost 12 years to make one revolution. This means that its eastwards motion among the stars is about 30° a year, or through one constellation of the zodiac. Although it never comes closer than 369 million miles (594 million km) to the earth, it is exceeded in planetary brightness only by Venus and occasionally Mars. To all outward appearances Jupiter is a gaseous body, and as such its rotation period varies in different latitudes; the equator spins at the fantastic rate of over 25 000 miles per hour (over 40 000 km/h) in 9 h 50 m, the equatorial bulging being a direct result of its high speed of rotation.

Despite its great distance from the earth, a telescope of moderate power will reveal quite a lot of detail in Jupiter's cloudy atmosphere. The principal markings run in belts or zones across the disc parallel to the planet's equator. The equatorial zone, usually very bright, is bordered to the north and south by other darker belts, within which numerous bright and dark spots are frequently revealed by larger telescopes and photographs. A very curious feature is a reddish, oval-shaped object drifting around in the upper atmosphere of the southern tropical belt as though it were a solid floating in a more liquid atmosphere below. It has been observed for over a century and has become known as Jupiter's Great Red Spot.

Jupiter is a very cold planet with surface temperatures in the region of —140°C. Certain gases are known to be present in the atmosphere, but because we can only see the outer layers the nature of Jupiter's interior is not known. One theory suggests that it has a rocky core surrounded by a thick layer of ice, the whole of which is enveloped by a dense outer atmosphere consisting of hydrogen, methane, ammonia and possibly helium. Another theory is that the planet is composed of about 80 per cent hydrogen, under such great pressure near the centre as to make it solid, with a slushy middle layer below the outer blanket of atmosphere. In any event Jupiter bears little resemblance to the earth or any of the other terrestrial planets. However, it is unique among the planets inasmuch as it is a strong emitter of radio waves. Investigations carried out since 1955 indicate that it possesses an ionosphere and also a magnetic field associated with intense zones of trapped solar radiation similar to the Van Allen belts surrounding the earth.

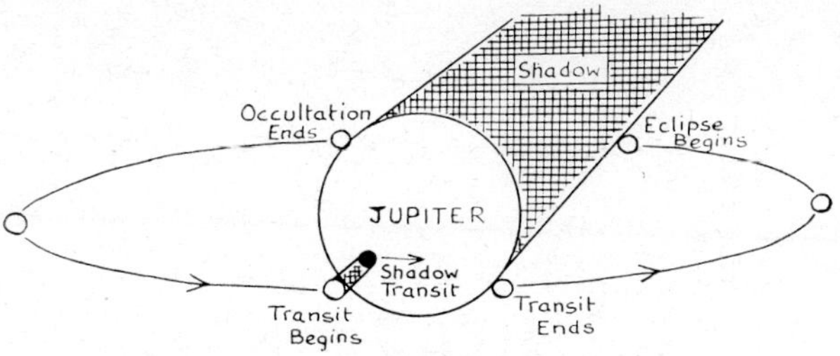

Fig. 6.9
Phenomena of a Satellite of Jupiter.

Jupiter has 12 satellites, but eight of them are so small that they can be detected only with large telescopes using photography. The other four, comparable in size to the moon, are bright enough to be seen through a small telescope or a good pair of binoculars. The bright satellites, Io, Europa, Ganymede and Callisto, were discovered by Galileo in 1610 and are sometimes referred to as the *Galilean satellites*. The Galileans are numbered 1 to 4, in order of distance from Jupiter; the others are numbered in order of their discovery, which accounts for the fact that the fifth satellite is closest to the parent planet. The orbits of the bright satellites are nearly in the plane of Jupiter's equator such that their movement takes them behind the planet and into its shadow, giving rise to occultations and eclipses respectively. They also transit across the face of Jupiter when, although drowned in its strong light, their shadows are cast upon the planet's disc. (fig. 6.9).

TABLE 6.2
JUPITER'S BRIGHT SATELLITES

Name	Dist. from Jupiter in diameters of Jupiter	Period of Revolution in Days	Diameter	
			Miles	Kilometres
Io	5.9	1.3	2000	3200
Europa	9.4	3.2	1810	2900
Ganymede	15.0	7.1	3125	5000
Callisto	26.4	16.3	2810	4500

SATURN

Outwards from the sun, the next planet beyond Jupiter is the "ringed planet" Saturn, at a mean distance of 887 million miles (1427 million km). Like Jupiter its fast rotation, 10 h 12 m, is evident from the polar flattening, or equatorial bulging, which is more pronounced than that of any other planet:

equatorial diameter	74 200 miles (119 500 km)
polar diameter	66 900 miles (107 700 km)

Its physical nature is also similar to Jupiter's—hydrogen, methane and ammonia—except that the surface temperature is some 30°C lower. Bands running parallel to the equator are visible through larger telescopes, as are occasional spots, though they are less clearly defined than those of Jupiter due to its much greater distance from us. To the naked eye Saturn appears as a yellowish first magnitude star, but even a low-powered telescope will reveal the planet's unique ring system which makes it one of the most splendid objects to be seen in the night sky. Galileo, who detected the rings in 1610, was unable to see their shape clearly in his small telescope, and nearly fifty years elapsed before Huygens interpreted them correctly. The system comprises three concentric rings: the *outer ring*, the middle or *bright ring*, and the inner or *crape ring*, with a distinct gap, known as Cassini's division, between the middle and inner rings. The overall diameter of the ring system is about $2\frac{1}{3}$ times the equatorial diameter of Saturn; its thickness is still a matter of some doubt, estimates ranging from several inches to several miles. Because the rings are inclined 27° to the plane of the planet's orbit, their northern and southern faces are tipped alternately towards us as shown in fig. 6.10. At intervals of about 15 years they appear edgewise on to us (last happened in 1966) and become invisible in all but the largest telescopes. The rings consist of a multitude of small ice-covered rocks, or ice particles perhaps no bigger than snowballs, revolving around Saturn in individual orbits. Their origin is not known, but one suggestion is that they are the remains of a satellite which approached the planet too closely and was ripped apart by tremendous gravitational forces. In 1848, the French astronomer Roche

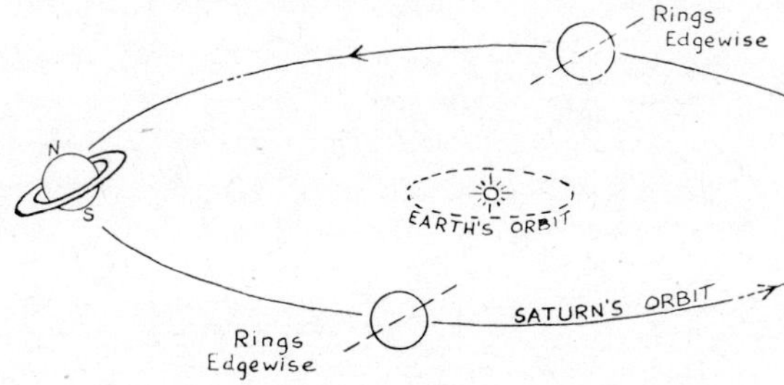

Fig. 6.10
The Changing Aspect of Saturn's Rings.

calculated that any satellite approaching to within $2\frac{1}{2}$ times the radius of the parent planet would disintegrate—the rings are within *Roche's limit* for Saturn. Saturn has 10 satellites, whose names in order of distance from the planet are:—

Janus, Mimas, Enceladus, Tethys, Dione, Rhea, Titan, Hyperion, Iapetus and Phoebe. In terms of Saturn's diameter Janus, discovered in 1966, lies near the outer edge of the rings at a distance of 1·3 diameters (just outside Roche's limit), whilst Phoebe is over 100 diameters away. The largest of the satellites, Titan, has a diameter of only a little less than half the earth's, and it is the only satellite in the solar system on which an atmosphere (methane) has been detected. The other satellites have diameters of less than 800 miles (1300 km), the smallest, Phoebe, being only 120 miles (200 km) across.

URANUS, NEPTUNE AND PLUTO

In order of distance from the sun, the three planets beyond Saturn are Uranus, Neptune and Pluto (for data see Table 6.1). Uranus and Neptune are two rather similar planets, intermediate in size between Earth and Jupiter, with physical natures probably much the same as those of Jupiter and Saturn except that they are very much colder. Whilst Uranus may just be visible to the naked eye as a 6th magnitude star, a telescope is required in order to find Neptune, two magnitudes fainter. At such great distances even the largest of telescopes fail to reveal any significant features on their small bluish-green discs, apart from a few faint bands on Uranus. A peculiar feature of Uranus is that it spins on its axis with the equator almost perpendicular to the plane of its orbit. This means that as it revolves first one pole, then the equator, then the other pole are presented to the sun, giving each pole a summer and winter each lasting 42 years. Uranus has five known satellites—Miranda, Ariel, Umbriel, Titania and Oberon—whereas Neptune has only two, Triton and Nereid.

The most distant planet, Pluto, with a diameter of less than half the earth's, is a mere dwarf compared to the other outer planets; photographed through the world's largest telescope it appears only as a faint star. Its orbit is highly inclined to the ecliptic (17°), and because of this it is not confined to the zodiac like the other planets. Another odd feature is the large eccentricity of its orbit which makes it possible for Pluto to come closer to the sun than Neptune, as it will do for about 20 years commencing in 1979. These peculiarities in Pluto's motion have led to the suggestion that it may not be a true planet at all, but an escaped satellite of Neptune. As to the suggestion that Pluto may not be the last planet of the solar system, it can only be said that so far there is no evidence to suggest the existence of another one further out.

The story of the discoveries of the three outer planets began in 1781, when Sir William Herschel in England found Uranus accidentally whilst making an examination of the stars in the constellation Gemini with a 6½-in (165 mm) telescope. Herschel, a German who fled to England from war-torn Europe, spent his early days as a musician but his great love of astronomy eventually led him to give up music completely and, despite having had little formal education, he became one of the greatest astronomers of his time. He made many large telescopes, including a 48-in (1·2 m) reflector, with which he carried out systematic surveys of the stars. With sister Caroline acting as his assistant, Herschel recorded several thousand double stars, nebulae and star clusters (see Chapter 7). His studies of the motions of double stars suggested to him that the law of gravitation acted outside the solar system. Not only did he obtain a reasonable idea of the shape of our own star system (the Galaxy), he also concluded that as a star the sun was of no special note and in fact moved in space towards the constellation Hercules. Because of his pioneer work in the study of the stars, Herschel is often called the Father of stellar astronomy.

After due allowance for the disturbing effects of the known

planets (perturbations), it later became clear that Uranus was not following the orbit computed for it, but was being drawn off course by a planet as yet unseen. Using the law of gravitation, Leverrier in France and Adams in England independently calculated the position of the unknown planet and, although Adams completed his calculations first, it befell Galle in 1846, acting upon Leverrier's calculations, to discover Neptune within one degree of the position predicted for it. Needless to say, the discovery of Neptune was hailed as a triumph for the law of gravitation.

The outermost planet, Pluto, was discovered photographically in 1930 by Tombaugh at the Lowell observatory during the continuation of a search started many years before by Lowell himself.

COMETS

Comets attract a good deal of attention because they emerge suddenly from the depths of space, often moving very rapidly against the starry background and frequently becoming bright enough to be seen with the naked eye. In days of old, the appearance of a bright comet was regarded as an omen of impending disaster such as war or pestilence, and there are many examples of artists' impressions in which comets are depicted as demons descending upon the earth. Of course, not all comets are spectacular; the majority are relatively faint objects, perhaps only about one in every five reported ever becoming visible to the naked eye. Professional astronomers frequently detect comets on photographs taken for other purposes, but many discoveries are made by amateur astronomers who undertake detailed searches of those areas of the sky where comets are most likely to be found—in the west after sunset, and in the east before sunrise. A new comet is initially designated by the year of its discovery followed by a letter in order of its discovery, though in some cases it becomes better known by the name(s) of the discoverer(s). For example, Comet Arend-Roland, the eighth comet to be discovered in 1956 was also 1956 h. In catalogues comets are given a permanent designation.

THE ORBITS OF COMETS

The planets revolve around the sun in a direct sense, from west to east, in orbits more or less in the same plane as that of the earth's so that they are usually to be found close to the ecliptic. On the other hand comets are not nearly so orderly in their movements, and apart from the fact that they revolve around the sun in accordance with the law of gravitation they show scant regard for the planetary "highway code". They may travel with either direct or retrograde motion in random orbits which are highly elongated and often steeply inclined to the ecliptic. As with a planet, the point in a comet's orbit where it is nearest the sun is called perihelion, the furthest point being known as aphelion. Since a comet can only be observed over the small portion of its orbit when it approaches the vicinity of the sun, it is not always possible to determine the complete orbit with any degree of accuracy. However, the orbits of comets do seem to fall into two main groups:

(1) Comets having nearly *parabolic* orbits. A parabola is an open curve which extends to an indefinite distance, the two sides becoming nearly parallel. The orbits of many comets are so nearly parabolic that their aphelions are far out beyond the orbits of the planets, with periods of revolution so large that only one appearance of each comet has ever been recorded.

(2) Comets having definitely elliptical orbits. Because they are capable of being seen on more than one return to perihelion, they are known as *periodic comets*. The giant planets, on account of their strong gravitational pull, are capable of drawing a comet

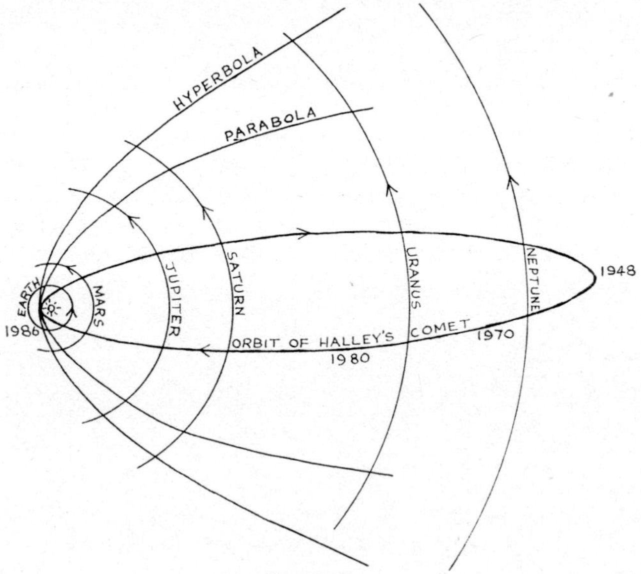

Fig. 6.11
The Orbit of Halley's Comet.

moving in a near-parabolic orbit off course and forcing it into an elliptical orbit to become a short-period comet. This "capture" process is very evident in the case of Jupiter which has a family of over two dozen comets whose aphelions are somewhere near the orbit of the planet. Encke's comet, a faint member of Jupiter's family, has the shortest period ($3\frac{1}{4}$ years) of any known comet. The most famous of the periodic comets, Halley's comet, has been seen on 30 occasions since 467 B.C. It was named after Edmond Halley, the second British Astronomer Royal, who calculated the orbit of the bright comet of 1682 and noted its close resemblance to the orbits that he had similarly calculated for the earlier comets of 1531 and 1607. He concluded that they were successive appearances of the same comet having a period of about 76 years, and duly announced that it would return in 1758. Although Halley died before the comet's next return, it was sighted according to his prediction. It was also seen in 1835 and 1910, and the next return is expected in 1986. Halley's comet (fig. 6.11) moves with retrograde motion in a highly elliptical orbit extending beyond the orbit of the planet Neptune which, like Jupiter, also possesses a small family of comets.

NATURE OF A COMET

A comet consists of a small *nucleus* of frozen material surrounded by a foggy envelope called the *coma*, accompanied by a *tail* which fans outwards away from the sun. The nucleus is of a porous nature composed mainly of ices such as methane, ammonia and water, in which are embedded quantities of meteoric material in the form of small particles of iron, nickel, calcium, silicon, sodium and other elements. When the comet is at great distances from the sun it does not possess a tail, but as it approaches perihelion the heat received from the sun causes some of the ices to evaporate and explode into the coma, carrying with them fragments of the meteoric material, which are finally expelled through the comet's tail or scattered along its orbit. Thus a periodic comet gets smaller after each return to perihelion so that after many returns it disappears, leaving the original solid material to revolve around the sun as a meteor stream. A comet becomes visible due partly to the glowing of the gases and partly on account of reflected sunlight. The tail may extend for many millions of miles into space, but because it is extremely tenuous it is readily repelled by solar radiation pressure and is consequently always directed away from the sun, (fig. 12). The average comet, often tailless, is usually too faint to be seen without some form of

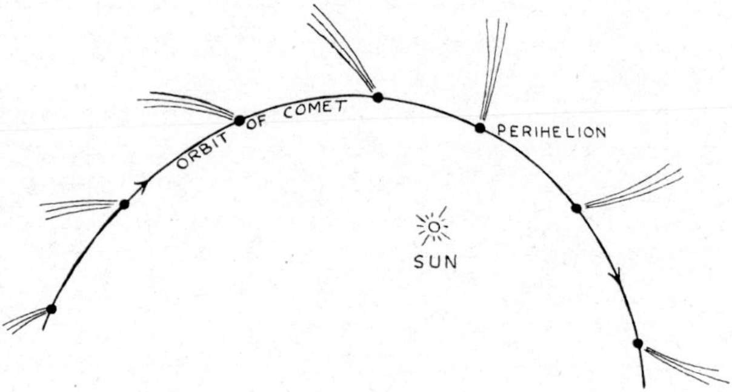

Fig. 6.12
The Tail of a Comet is Directed Away from the Sun.

optical aid, but every now and then the heavens are graced by a really spectacular one which attracts general attention. Comet Arend-Roland, which appeared in 1956, developed a tail well enough to be seen in the night sky with the naked eye. Comet Ikeya-Seki, the Great Comet of 1965, became so brilliant that it could be seen with the naked eye in broad daylight. Its tail grew to over three-quarters of the earth-sun distance in length, providing a truly magnificent object to observers in lower northern latitudes and the southern hemisphere. Comet Bennett, 1969*i*, seen first in the southern sky, sped northwards through the heavens with a tail of over 10° in length, giving northern observers their most spectacular sight since the last return of Halley's comet.

METEORS AND METEOR STREAMS

Meteors are particles of space material which revolve around the sun. They become visible only when they enter the earth's atmosphere and burn up due to friction with the air, producing short-lived luminous trails across the sky. They are often called "shooting stars" or "falling stars", but these names are totally misleading as meteors have no connection whatever with stars.

NATURE OF METEORS

Meteors are very small stony and metallic particles, mostly no bigger than grains of sand. They start to glow when they are about 80 miles (130 km) in height, the majority completely burning up by the time they are 50 miles (80 km) from the earth's surface. Larger meteors become visible at greater heights and come closer to the ground. Unusually large ones sometimes break up into smaller pieces, accompanied by explosions, their greater brightness earning them the name *fireballs*.

In addition to *sporadic meteors*, those which may occur at any time, there are *streams* of meteors revolving around the sun which give rise to *meteor showers* when they encounter the earth's atmosphere. The meteors in a stream move in nearly parallel paths and the trails they leave in the sky are also parallel although they may not appear to be so. But, just as railway lines seem to diverge from a distant point, so the meteor trails appear to emanate from a point in the sky. This point is known as the *radiant*, and meteors are often named after the constellation in which the radiant is situated (fig. 6.13). Thus the radiant of the Leonids is

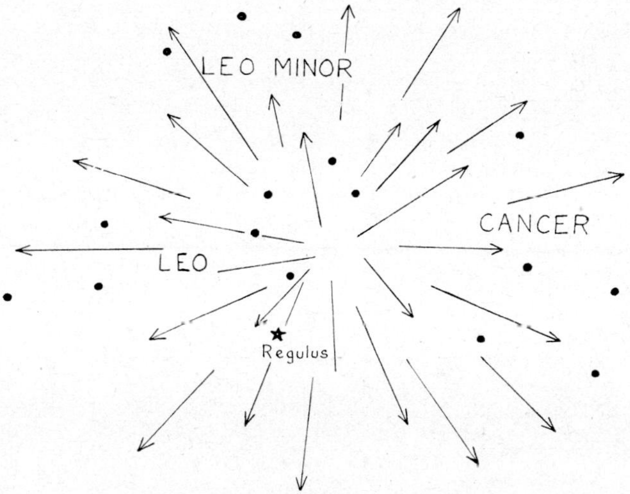

Fig. 6.13
Meteors Appear to Diverge from the Radiant.

in Leo, the Geminids in Gemini, and the Taurids in Taurus. The orbits of many meteor streams cut across the earth's orbit so that the earth, as it revolves around the sun, encounters each stream at a particular time of year (see Table 6.7). The most noteworthy shower of recent times occurred in 1966, when the Leonids gave such an intense display that it was almost impossible to count their numbers accurately, some observers estimating a maximum rate of over a thousand a minute. Unfortunately, although they occur annually, the Leonids only give spectacular displays every 33 years.

As mentioned earlier, the meteoric material contained in the head of a comet is expelled through the comet's tail and distributed along the orbit. There are several recorded cases of the dispersal of a comet into a meteor stream the most remarkable of which is that of Biela's comet. After its discovery in 1826 Comet Biela reappeared about every $6\frac{1}{2}$ years. At the 1845 return it was observed to split in two, and the double comet was seen again in 1852. On the following two predicted returns nothing was seen, but in 1872 instead of the comet there was a magnificent meteor shower—the Andromedids or Bielids. Further showers occurred in 1885 and 1898, but the stream has now practically disappeared.

METEORITES

A *meteorite* is a mass of stony or metallic space material which survives its passage through the earth's atmosphere and reaches the ground. The fall of a meteorite is often quite spectacular, with bright trails in the sky persisting for several minutes together with flashes of light and explosive effects. Meteoric material is frequently found scattered over a wide area, either as the result of the entry into the earth's atmosphere of a compact group of bodies or the disintegration of a larger single body. In the present century there have been two large falls in Siberia: the first in 1908, of which no trace has been found, devastated a large area of forest; the second in 1947, broke up causing many small craters. Sometimes a meteorite ricochets off the ground and finally comes to rest on the surface, like the 70 ton Hoba, South-West Africa meteorite, the largest known meteorite in the world. Another large one weighing 34 tons, discovered in Greenland by the famous polar explorer Peary, is on display at the Hayden planetarium in New York. Occasionally meteorites gouge out large craters in the earth's surface, over a dozen in various parts of the world having been recognised as of meteoric origin. The largest of these meteorite craters, situated near Canyon Diablo in Arizona, is a circular depression 4200 ft (1280 m) across and 570 ft (175 m) deep. Meteorites can occur at any time, their spasmodic nature indicating that they are not connected with meteor streams. One theory suggests that they stem from the collision of asteroids.

REVIEW QUESTIONS

1. Name the points in a planet's orbit where it is furthest from and nearest to the sun. At which of these points does a planet attain its greatest orbital speed?

2. The mean distance of the earth from the sun is known as an astronomical unit. If the sidereal periods of revolution of the Earth and Venus are respectively $365\frac{1}{4}$ days and 225 days, find by means of Kepler's third law the mean distance of Venus from the sun in astronomical units.

3. State Newton's law of gravitation. How would the force between two equal masses be affected (a) if each mass is doubled? (b) if the masses are unaltered but the distance between them is doubled?

4. Name two astronomers who have each discovered four planetary satellites.

5. Give the names of the following: (a) the first known periodic comet, (b) the comet having the shortest period, (c) the Great Comet of 1965.

6. What is a meteor radiant? Why are the Leonid and Bielid meteor showers so named?

7. State which of the following statements are true or false: (a) The synodic period of a planet is greater than its sidereal period. (b) A superior planet can be observed in crescent phase from the earth. (c) An inferior planet transits across the face of the sun at every inferior conjunction. (d) A superior planet can cross an observer's meridian at local midnight.

8. The period of Halley's comet, last seen in 1910, is just over $76\frac{1}{2}$ years. Could the bright comet seen around the time of the Battle of Hastings in 1066 have been Halley's comet?

PROJECTS AND RESEARCH

1. THE PATHS OF THE PLANETS

Since the planets move with respect to the stars, the best means of telling where they are at any time of the year is to draw their paths on a star chart similar in construction to the one described in Project 4, Chapter 2. If a separate chart is to be made specially for the purpose the declination scale need not extend beyond 35° N and S, that is just sufficient to include the ecliptic and the zodiacal limits. When the chart is ready, the next thing to do is to extract the co-ordinates of the planets from Whitaker's Almanack, or other available source, and plot the values of Right Ascension and declination at convenient intervals, joining up the points to give the smoothest possible curve in each case. Neptune and Pluto may be omitted as they cannot be readily observed. Where an inferior planet comes to inferior conjunction, or a superior planet to opposition, during the year in question, the retrograde loop will show clearly on the plot. It will also be noticed that the further a superior planet is from the sun the shorter is its annual track: the path of Mars will run about half way across the map, that of Jupiter through about two hours of R.A., that of Saturn even less, and that of Uranus less again. In addition to showing the positions of the planets at all times, the completed diagram can be used to find out if any planets are likely to be in the same region of the sky at the same time. Inserting the sun's ecliptic position on the map will reveal when a planet is furthest from the sun in the sky and thus reasonably placed for observation. Similarly, plotting the moon's path for any given month will indicate when either of the planets will appear close to the moon.

2. OBSERVING THE PLANETS

With the exception of Neptune and Pluto, all the major planets are capable of being seen with the naked eye. If the preceding project has been tackled it will be evident that the proximity of the planets to the ecliptic and their movements relative to the constellations is sufficient to distinguish them from stars. Also, bright planets can often be recognised by their steady light when the stars around them are twinkling, or scintillating. The reason for this is that to us a star is a point source, the light from which is often seriously disturbed as it passes through the earth's atmosphere, whereas a planet appears larger and the twinkling effects from different parts of the disc tend to cancel out. Nevertheless, near the horizon, where the light has to pass through a greater thickness of atmosphere, even planets can twinkle. Of course, Venus and Jupiter are so brilliant that there is no chance of confusing either of them with stars. Mars, brightest at opposition, is easily distinguished by its red colour. Mercury is also easily recognised as a planet when seen, though its location presents something of a challenge. Because it is never too far from the ecliptic the search for Mercury is more likely to be fruitful when maximum eastern or western elongation (evening and morning star respectively) coincides with the steepest inclination of the ecliptic to the horizon. In middle and high northern latitudes the planet is to be found low in the west after sunset at a spring eastern elongation, and in the east before sunrise at an autumn western elongation. In the tropics the ecliptic is more nearly perpendicular to the horizon at most times of year, so a morning or evening sighting is a much better proposition. In southern Australia, New Zealand and South Africa, the most favourable opportunities are when the planet is

TABLE 6.3
VISIBILITY OF THE PLANETS

PLANET	NAKED EYE	LARGE BINOCULARS OR 2-in (51 mm) TELESCOPE	LARGER TELESCOPES UP TO 4-in (102 mm)
Mercury	Planet only Pinkish-white	Planet only Perhaps phase	Distinct phase No surface features
Venus	Planet only Very bright White	Distinct phase Best observed at twilight	Phase, vague surface shadings at twilight
Mars	Planet only Brightest at opposition Reddish	Definite disc	Larger disc, surface features—polar caps and dark areas
Jupiter	Planet only Very bright White	Definite disc Possibly faint surface bands Bright satellites	Larger disc Polar flattening Surface bands and satellites
Saturn	Planet only Yellowish	Probably oval-shaped	Ring system Faint surface bands Polar flattening
Uranus	Starlike, faint	Starlike, bluish	Small bluish-green disc

TABLE 6.4

FORTHCOMING ELONGATIONS OF MERCURY

YEAR	EASTERN ELONGATION	WESTERN ELONGATION
1970	April 18 August 16 December 10	February 5 June 5 September 28
1971	April 1 July 29 November 23	January 19 May 17 September 12
1972	March 14 July 10 November 5 —	January 1 April 28 August 25 December 14
1973	February 25 June 22 October 18	April 10 August 8 November 27
1974	February 9 June 4 October 1	March 23 July 22 November 10
1975	January 23 May 16 September 13	March 6 July 4 October 25
1976	January 7 April 28 August 26 December 20	February 16 June 15 October 7 —
1977	April 10 August 8 December 3	January 28 May 27 September 21
1978	March 24 July 22 November 16 —	January 11 May 9 September 4 December 24
1979	March 8 July 3 October 29	April 21 August 19 December 7
1980	February 19 June 14 October 11	April 2 August 1 November 19

TABLE 6.4 (continued)

YEAR	EASTERN ELONGATION	WESTERN ELONGATION
1981	February 2 May 27 September 23	March 16 July 14 November 3
1982	January 16 May 8 September 6 December 30	February 26 June 26 October 17 —
1983	April 21 August 19 December 13	February 8 June 8 October 1
1984	April 3 July 31 November 25	January 22 May 19 September 14
1985	March 17 July 14 November 8 —	January 3 May 1 August 28 December 17

a morning star in southern autumn and an evening star in southern spring. Saturn can be confused with a star, but once it has been identified it is easy to keep track of on account of its rather leisurely movement among the stars. Uranus is not easy to find unless its position is first marked on a star map. Naked-eye observations of the planets will not reveal any surface features; identification and positional observations are about all that can be accomplished. But these can be quite interesting, particularly when a planet is closest to the earth and performing its retrograde motion, and a certain amount of satisfaction can be gained from checking a planet's progress among the stars. The amount of detail capable of being seen with binoculars or a small telescope depends on several factors, not least the optical quality of the instrument. Table 6.3 gives some idea of what to expect.

3. DAYLIGHT SIGHTING OF VENUS

Given a clear blue sky with no hint of haze, it is possible to locate Venus in the daytime with the naked eye. When the planet is a conspicuous morning or evening star it is far enough from the

TABLE 6.5
FORTHCOMING ELONGATIONS OF VENUS

YEAR	EASTERN ELONGATION	WESTERN ELONGATION
1970	September 1	—
1971	—	January 20
1972	April 8	August 27
1973	November 13	—
1974	—	April 4
1975	June 18	November 7
1976	—	—
1977	January 24	June 15
1978	August 29	—
1979	—	January 18
1980	April 5	August 24
1981	November 11	—
1982	—	April 1
1983	June 16	November 4
1984	—	—
1985	January 22	June 13

TABLE 6.6
FORTHCOMING OPPOSITIONS OF MARS

YEAR	OPPOSITION	CONSTELLATION
1971	August 10	Capricornus
1973	October 25	Pisces-Aries
1975	December 15	Taurus
1978	January 22	Gemini-Cancer
1980	February 25	Leo
1982	March 31	Virgo
1984	May 11	Libra
1986	July 10	Sagittarius

sun in the sky to make a daylight sighting more likely. It helps to know what one is looking for of course, and Venus appears as a very small bright white dot—the author has used daylight observations of the planet on countless occasions for the purpose of finding position when at sea. In the clear skies of the tropics finding the planet is a relatively simple task, but in the temperate latitudes it is a little less easy because the planet does not rise nearly so high in the sky. The best plan is to wait until Venus is on the meridian, when its altitude is greatest, and set up a straight-edge in conjunction with a protractor and a plumb line to the altitude at meridian passage. Sighting along the straight-edge will then direct one's gaze to the region of the sky in which the planet is situated. If the Right Ascension and declination of the planet are known, the method for finding the meridian passage time and altitude is set out in the example given below. Some almanacs provide the meridian passage time, in which case the first part of the calculation is not necessary.

Suppose an observer at Cardiff (lat. 51° 30′ N, long. 3° 15′ W) requires the time and altitude of Venus (R.A. 0 h 56 m, decl. +5°·8) at meridian passage on 16 May 1969.

From the Sun Table (Chapter 1), or other source, the sun's R.A. on the same date is found to be 03 h 30 m, and the Equation of Time —4 m.

Since the R.A. of Venus is less than that of the Sun, Venus crosses the meridian before the sun by a time interval equal to the difference of their R.A's.

Local apparent time of sun's transit		12 h 00 m (always)
Equation of time	—	4 m
Local mean time of sun's transit		11 h 56 m
Difference of R.A.'s	—	2 h 34 m
Local mean time of Venus' transit		09 h 22 m
Longitude west in time	+	13 m
GMT (or UT) of Venus' transit		09 h 35 m

Observers in Britain need to add one hour for BST, thus giving a meridian transit time of 10 h 35 m (1035 A.M.). Had the R.A. of Venus been greater than that of the sun, Venus would have crossed the meridian after the sun.

The altitude calculation is similar to that for a star as described in Project 2, Chapter 2.

latitude	51°·5 north	same names—subtract
declination	5°·8 north	
zenith distance	45°·7	
	90°·0	
altitude	44°·3	

If a pair of binoculars or a small telescope is set up rigidly in the meridian to the required altitude, the planet should be in the field of view close to the calculated time.

N.B. The above methods can be used to calculate the time of meridian passage and meridian altitude of other planets, or any astronomical body whose R.A. and declination is known. (For meridian passage of a star, see Chapter 2 and Appendix I.)

4. OBSERVING METEORS

From any place well away from the glare of town lights it would be quite normal for a patient observer to see a dozen or so sporadic meteors streak across the sky on any night of the year. Their frequency is likely to be greater after midnight for the simple reason that during the evening we are on the following side of the earth as it moves in its orbit, so that we are protected except from the meteors that overtake us; in the morning we are on the leading side of the earth, exposed to the incoming meteors. Of course, meteors are much more plentiful when the earth passes through one of the meteor streams encountered at fairly regular

TABLE 6.7
MAIN ANNUAL METEOR SHOWERS

NAME	DATE	REMARKS
Quadrantids	Jan. 3-5	Very sharp maximum on Jan. 4. Radiant in Draco-Bootis region.
Lyrids	April 19-22	Swift meteors.
Eta Aquarids	May 1-8	Swift meteors with long paths. Maximum May 5-6.
Delta Aquarids	July 25-30	Slow meteors.
Perseids	July 27-Aug. 17	Rich shower of swift meteors. Maximum Aug. 11-12.
Orionids	Oct. 15-25	Swift meteors. Maximum Oct. 20.
Taurids	Oct. 26-Nov. 16	Slow meteors. Maximum Nov. 1-9.
Leonids	Nov. 14-17	Unpredictable shower of swift meteors.
Geminids	Dec. 9-14	Rich shower. Maximum Dec. 13.
Ursids	Dec. 20-22	Not always plentiful. Radiant in Ursa Minor.

times each year. Some of the showers worth looking for are listed in Table 6.7. It will be recalled that shower meteors emanate from a radiant, and the number of trails seen depends to a large extent on where the radiant happens to be at the critical time, fewer meteors being seen when it is low in the sky than when it is nearer to the zenith. Another factor to be considered is the moon, for its presence in the night sky will drastically reduce the chances of seeing a good display of meteors.

5. THE ZODIACAL LIGHT

On a dark clear night a faint glow known as the *Zodiacal Light* may be seen in the west after dusk and in the east before dawn. It is due to sunlight scattered by a ring of minute dust particles around the sun in the plane of the solar system. The Zodiacal Light, so named because it lies very close to the ecliptic, takes the form of a narrow cone, broadest and brightest near the horizon where its width may be as much as 25°, tapering upwards to an apex about 90° from the sun's position below the horizon. Its proximity to the ecliptic makes its faint glow best seen when the ecliptic is most steeply inclined to the horizon, as in spring evenings and autumn mornings (see Project 2). In the tropics, the ghostly glow of the Zodiacal Light spreading almost vertically into the sky can be seen most mornings and evenings through the year.

6. THE PLANETARY SYSTEM

Draw up an account of the sun's family of planets: compare their sizes, distances from the sun, physical natures, atmospheres, possibility of life, etc. Mention a unique feature of each planet—Mercury has no atmosphere; a day on Venus is longer than a year; and so on. Include in your account photographs taken from scientific magazines, together with a full description of any planetary observations carried out. In order to gain some appreciation of the scale of the solar system, draw the orbits of the planets in the form of circles or radii equal to their mean distances from the sun on a map of the country, taking the capital city or your town as the position of the sun; or on a local map, placing the sun at the Civic Centre.

7. THE EARLY ASTRONOMERS

Research into the work of Eratosthenes, Hipparchus, Ptolemy and other ancient astronomers. Examine the validity of their conclusions in the light of present knowledge.

8. THE MODERN ASTRONOMERS

The foundations of modern astronomy laid by Copernicus in the 16th century were subsequently built upon by Galileo, Tycho Brahe, Kepler, Isaac Newton, William Herschel and others. Research into the lives of these great men: find out more about their work, the times in which they lived and the prejudices they encountered when their ideas conflicted with traditional teachings. Flamsteed, Halley and Bradley were the first in the line of distinguished British Astronomers Royal—find out about them too, extending your reading to scientific magazines with the object of discovering something about the eminent astronomers of today.

9. COMETS

Comet-hunting can hardly be rated as an easy project to undertake; it is more the speciality of the devoted amateur astronomer who will spend hundreds of hours observing over many years, with perhaps no reward for his labours except for an excellent working knowledge of the night sky. Also, the equipment required is rather more than the reader of this book is assumed to possess. Binoculars of at least 2-in aperture mounted on a firm stand are a prerequisite, together with a detailed star atlas and an unlimited amount of patience. Short of joining the small fanatical band of comet seekers, the best one can do is look for known comets. The Handbook of the British Astronomical Association (BAA) contains details of periodic comets, and members are notified of new discoveries in the Association's circulars. Anyone intent on swelling the ranks of the amateur astronomers is well advised to join the BAA.

We have seen that comets follow either highly elliptical or near parabolic orbits; some are even thought to travel in near *hyperbolic* orbits (a hyperbola is a more open type of curve than a parabola, its branches approaching diverging straight lines). However, to the non-mathematically minded, words like ellipse, parabola and hyperbola may seem quite meaningless, yet it is not difficult to construct these curves geometrically. The method of drawing an ellipse has already been described in (Project 7, Chapter 3) but the experiment should now be repeated, increasing the distance between the two pins until really elongated ellipses are made.

To construct a parabola the following materials are required: a large sheet of paper, a straight edge, a large set square, a pencil, pins, a length of string, and preferably a flat board to avoid making holes in the table. The method is illustrated in fig. 6.14.

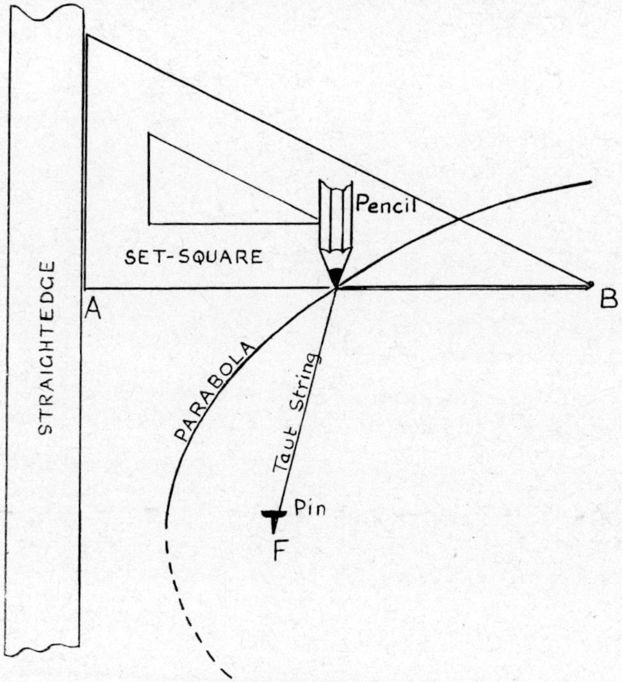

Fig. 6.14
Construction of a Parabola.

Lay the sheet of paper on the board, securing the straight edge firmly upon it. A string of length AB is secured to the point B of the set square and to the point F, the focus of the parabola. Keeping the string stretched tightly by means of the pencil pressed hard against the edge of the set-square, move the set-square along the straight edge and one side of the parabola will be traced out. On reaching the limit of travel, rearrange things on the other side of the pin and draw the other side of the parabola.

In the construction of a hyperbola, the point P of the pencil is fastened to a string that slides smoothly around two pins F_1 and F_2 (fig. 6.15). The pencil must not slide along the string as when drawing the parabola, and to prevent such movement make a couple of loops of the string around the pencil point. It is also necessary to ensure that PF_2 exceeds PF_1 by an amount less than the distance between the pins. Now, keeping the string stretched tightly, pull on the free ends together and one side of the hyperbola will be drawn. Then, with the pencil on the other side of the pins, draw the other side of the hyperbola by paying out the string, making sure the string is tight and not slipping around the pencil.

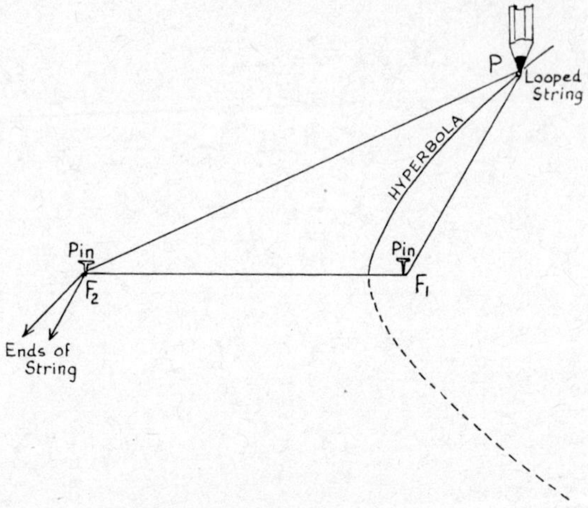

Fig. 6.15
Construction of a Hyperbola.

If a piece of wood in the form of an upright circular cone is available four cuts made as follows will yield the conic sections—circle, ellipse, parabola and hyperbola:

(1) A cut parallel to the base gives a circle.
(2) A cut not parallel to the base, but not passing through the base, gives an ellipse.
(3) A cut parallel to the edge of the cone gives a parabola.
(4) A cut at a smaller angle to the axis of the cone than that for a parabola gives a hyperbola.

10. PHOTOGRAPHY

Without rather elaborate equipment detailed planetary photography is not a feasible proposition, but this does not mean that nothing useful can be done with just an ordinary camera. For example, if one is tracking the path of a planet, direct photographs taken from night to night provide an interesting record. Exposures of a few seconds are quite sufficient to give photographs from which the position of the planet relative to the nearby stars can be measured. Also, the close approach of a planet to the moon or another planet is well worth a shot, long exposures being given if trails are desired (see projects on photography in Chapters 1, 2 and 5).

TABLE 6.8

HIGHLIGHTS OF PLANETARY EXPLORATION

SPACECRAFT	LAUNCH DATE	COUNTRY	REMARKS
	1960		
Pioneer 5	Mar. 11	USA	Deep space probe towards Venus. Measured interplanetary magnetic fields, micrometeorites, etc.
	1961		
Venus 1	Feb. 12	USSR	Passed 62 000 miles from Venus. Communications failed.
	1962		
Mariner 2	Aug. 27	USA	Passed 21 000 miles from Venus. Measured interplanetary magnetic fields and cloud temperatures on Venus.
Mars 1	Nov. 1	USSR	Passed 120 000 miles from Mars. Communications failed.
	1964		
Zond 1	April 2	USSR	Deep space probe towards Venus. Communications failed.
Mariner 3	Nov. 5	USA	Mars probe. Communications failed.
Mariner 4	Nov. 28	USA	Passed 6120 miles from Mars. Sent back photographs of Martian surface.
Zond 2	Nov. 30	USSR	Mars probe. Communications failed.
	1965		
Zond 3	July 18	USSR	Deep space probe. Photographed far side of the moon.
Venus 2	Nov. 12	USSR	Passed 15 000 miles from Venus. Communications failed.
Venus 3	Nov. 16	USSR	Crash landed on Venus.
Pioneer 6	Dec. 12	USA	Similar to Pioneer 5.
	1966		
Pioneer 7	Aug. 17	USA	Similar to Pioneers 5 and 6.
	1967		
Venus 4	June 12	USSR	Soft landed on Venus. Measured planetary temperatures.
Mariner 5	June 14	USA	Passed 2600 miles from Venus.
Pioneer 8	Dec. 13	USA	Similar to earlier Pioneers.
	1969		
Venus 5	Jan. 5	USSR	Entered Venus' atmosphere.
Venus 6	Jan. 10	USSR	Entered Venus' atmosphere.
Mariner 6	Feb. 24	USA	Sent back 74 photographs of Mars.
Mariner 7	Mar. 27	USA	Sent back 126 photographs of Mars.

11. SPACE PROBES

Using the information provided in Table 6.8 as a basis for research, prepare a detailed account of the exploration of the planets Venus and Mars by unmanned spacecraft.

ANSWERS TO REVIEW QUESTIONS

1. Aphelion, perihelion. Perihelion.
2. 0·72 AU.
3. (a) quadrupled, (b) quartered.
4. Galileo and Herschel.
5. (a) Halley's comet, (b) Encke's comet, (c) Comet Ikeya-Seki.
7. (a) true, (b) false, (c) false, (d) true.
8. Yes.

CHAPTER 7

NATURE OF THE STARS

Stars are globes of intensely hot gases, but on account of their immense distances from us those we see in the night sky appear as mere point sources of light. The sun is a star too, seemingly larger than the other stars simply because it is very much closer to us. The several thousand naked-eye stars and the sun belong to the Galaxy, a gigantic spiral system containing somewhere in the order of 100 000 million stars. As far as we are concerned the sun is a most important astronomical body—we depend upon its light and heat for our very existence—but within the Galaxy it does not occupy a central position and is in most respects a fairly modest star. Beyond our own Galaxy, other galaxies are scattered like islands through the depths of space for as far as the largest telescopes can penetrate. These external galaxies are so remote that the light from the nearest of them takes about two million years to reach us.

THE SUN

The sun is the star about which the earth and other planets revolve. Of all the countless millions of stars in the universe, the sun is the only one whose surface can be observed. Thus astronomers pay it close attention, for the knowledge gained about "our star" leads to a better understanding of the nature and behaviour of stars in general.

STRUCTURE OF THE SUN

The sun is a hot, glowing sphere of gases held together by gravitational attraction. Its main constituent is hydrogen, and the energy of the sun is created deep in its interior by a conversion of the hydrogen into helium by thermonuclear processes. The extremely violent reactions going on inside the sun force energy outwards, preventing the sun from collapsing under its own gravity. The balance of forces is such that the sun is regarded as a fairly stable type of star, its hydrogen conversion proceeding at a pace which indicates a life-expectancy of billions of years. Many other elements are found in the sun, including nearly all of those known on earth; carbon, nitrogen, oxygen, sodium, magnesium, calcium and iron, to name a few. The temperature near the sun's centre is estimated at 15 000 000°C.

From the earth the sun, preferably viewed through a piece of smoked glass or a dark filter, presents a disc having an apparent diameter of about ½°; its real diameter is 864 400 miles (1 392 000 km). The solar disc we normally see, representing the deepest layer to which our vision can penetrate, is known as the *photosphere*. The temperature in this region is about 6000°C. Photographs of the photosphere reveal a fine structure in the nature of small bright patches, known as *granulation*. The bright granules, separated by darker, narrow channels of cooler gases, are in a state of continuous change and are thought to be the tops of convection currents welling up from the violent solar interior. The photosphere is also the region where *sunspots* occur.

Rising above the photosphere to a height of several thousand miles is the *chromosphere*, so named because of its reddish colour when seen during the total phase of a solar eclipse. Photographs show that myriads of short-lived spicules of hot gases keep emerging from its upper surface, giving it a grass-like appearance. Between the photosphere and the chromosphere is a shallow layer of cooler gases called the *reversing layer*; within the chromosphere itself the temperature rises gradually with increase of height to about 30 000°C.

The outer atmosphere of the sun extending for several million miles beyond the chromosphere is known as the *corona*. It is composed of very tenuous gases at high temperatures (millions of degrees) and minute dust particles, reflected light from which gives rise to the Zodiacal Light mentioned in Chapter 6. Near to the time of maximum sunspot activity the corona, with its delicate streamers, extends about equal distances in all directions; near sunspot minimum, coronal streamers extend further out from the equatorial regions than elsewhere. Fiery tongues of hot gases called *prominences* emerge from the chromosphere and spread into the corona, their reddish glow contrasting sharply with the feeble white light of the corona during a total eclipse of the sun. Prominences assume a variety of forms: while some protrude as columns showing little or no change, others are violently eruptive and motion pictures of them are particularly awe-inspiring. Using special equipment, astronomers are able to study the chromosphere, the prominences and the corona at times other than during total eclipses of the sun.

SUNSPOTS AND THE SUN'S ROTATION

Sunspots are temporary dark spots on the sun's disc, ranging in size from small specks to massive patches that are visible to the naked eye. Their existence was known to early astronomers, centuries before the invention of the telescope, but systematic records of their appearance date only from 1610, when Galileo used his newly constructed instrument to observe the sun.

Small dark *pores* frequently develop in the photosphere, most of which disappear in a day or so, but sometimes several of these pores unite to form a sunspot. A well-developed spot consists of two distinct parts: a dark central area called the *umbra*, surrounded by a brighter region called the *penumbra*. The temperature of a sunspot is about 4500°C, or some 1500°C cooler than the photosphere, its dark appearance being simply an effect of contrast with the brighter surroundings. Small spots may last for a day or two; larger ones for longer. Often, they occur in groups, surviving for perhaps two weeks or more, a typical group comprising two fairly well-developed spots or pores of short duration.

Successive observations of spots show that they traverse steadily across the disc of the sun, thus providing evidence of its rotation. However, unlike the earth which rotates as a rigid body, the sun is gaseous and although the sense of its rotation is similar to that of the earth's (counter-clockwise viewed from above the north pole) the period of rotation is not the same in all latitudes. At the equator it is 25 days, at latitude 45° nearly 28 days, and at latitude 75° about 33 days.

Spots are carried around by the sun's rotation towards the western limb, *i.e.* the leading edge of the sun's disc as it moves westwards across the sky. Of the two main spots in a group, the leader is mostly the smaller and longer lasting. While the leader tends to assume a compact circular shape, the following spot spreads out in an irregular fashion, trailing towards the east. Before disappearing from the face of the sun, the larger spots break up into a number of smaller spots and pores.

The number of spots on the sun's surface varies considerably: there are times when its disc is completely devoid of spots and others when spots are present in great numbers. Records show clearly that there is a periodic rise and fall in sunspot activity about every 11 years, averaging about $4\frac{1}{2}$ years in the rise to maximum and about $6\frac{1}{2}$ years in the fall to minimum. Following a time of sunspot minimum, the first spots of a new cycle are seen around solar latitudes 30° N and S. As the cycle proceeds, the spot zones move closer and closer to the equator, being in about latitudes 15° at the times of maximum activity. By the end of the cycle the zones are in latitudes 5° N and S, and around this time the spots of the next cycle begin to appear in higher latitudes. Sunspots are rarely seen near the solar equator or in latitudes higher than 40°. The last sunspot maximum was in 1969.

An interesting aspect of sunspots is their association with strong magnetic fields. When the northern hemisphere leader has north polarity, the follower has south polarity—the reverse is true in the southern hemisphere. The polarities of the leaders and followers in each hemisphere are also reversed with each new cycle, so that there is a magnetic cycle having a period of 22 years.

SOLAR FLARES—EFFECTS ON THE EARTH

A *solar flare* appears as a sudden increase in brightness of the photosphere, often in the vicinity of a sunspot group, which builds up to a maximum intensity in a matter of minutes and dies away in half an hour or so. The violently eruptive nature of a flare results in the emission of ultraviolet radiation, together with spewing out of electrified particles which cause surges in the solar wind vastly in excess of that associated with the *quiet sun*. (Reference to the sections in Chapter 3 dealing with the earth's atmosphere and magnetic field may be found useful at this point.) The strong burst of ultraviolet radiation associated with a solar flare travels with the speed of light, reaching the earth in about eight minutes. The radio reflecting layers in the earth's ionosphere become agitated by the ultraviolet light, resulting in a partial or complete disruption of worldwide radio communications. The electrified particles travel much slower, arriving at the earth a day or so after the flare is seen. They result in widespread disturbances in the earth's magnetic field called magnetic storms, which are evidenced by the erratic behaviour of compass needles. Another striking effect resulting from solar flares are the colourful displays of *aurora* in the earth's upper atmosphere. The electrified particles are guided by the earth's magnetic field towards the polar regions, where collision with the particles of atmospheric gases releases energy characterised by a variety of colours, depending upon which gases have been excited. Aurorae occur more frequently in latitudes higher than 40° N and S. In the northern hemisphere they are known as *Aurora Borealis* or "Northern Lights", and in the southern hemisphere as *Aurora Australis* or "Southern Lights".

The connection between solar flares and sunspots indicates that the atmospheric disturbances described above also tend to follow an 11 year cycle.

STARS IN GENERAL

Having dealt with the sun in some detail, we can now turn our attention to other stars. How far away are they? Are they hotter or cooler than the sun? Are they bigger or smaller than the sun? In seeking the answers to such questions astronomers have found considerable variety among the stars.

DISTANCES OF THE STARS

The nearest star beyond the solar system is about 26 000 000 000 000 miles (42 000 000 000 000 kilometres) away from us. In order to avoid such cumbersome figures, astronomers use a larger unit of distance called the *parsec*, so named because it is derived from the words *parallax* and *second*.

Parallax is the apparent displacement of nearer stars with respect to those more distant, caused by the changing position of the earth as it revolves around the sun. A simple experiment may prove useful: whilst holding a finger vertically between your eyes, close and open each eye alternately and notice how your finger appears to move from side to side against the more distant background—this is parallax. Now move your finger further away from your eyes whilst keeping up the "winking" operation, and you will see that parallax decreases as the distance is increased. Fig. 7.1 shows the earth at two opposite points in its orbit, E_1 and E_2, such that the angle at the nearer star subtended by the diameter of the earth's orbit is x; parallax is half this angle. When

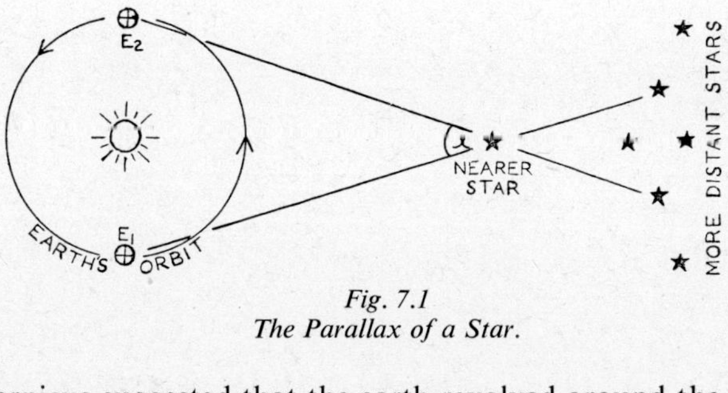

Fig. 7.1
The Parallax of a Star.

Copernicus suggested that the earth revolved around the sun, the main objection to his theory was that parallax could not be detected. At that time, the stars were thought to be much closer than they really are, and it was argued that if Copernicus was right then it should be possible to observe parallax readily; in fact stellar parallaxes are extremely small. Although Bradley proved the earth's revolution from the *aberration of starlight* in 1727 (Chapter 3), it was not until 1838 that Bessel succeeded in making the first measurement of parallax; that of the star 61 Cygni. This measurement, which amounted to only 0·3 second of arc, gave the first real indication of stellar distances, and also provided further convincing evidence of the earth's revolution. Nowadays, parallax is determined by comparing a series of photographs of the stars taken at six-monthly intervals.

The parsec is defined as the distance at which a star would have a parallax of one second. Since parallax decreases with increasing distance, then:

$$\textit{distance in parsecs} = \frac{1}{p}$$

where p is the parallax in seconds.

A more popular unit of astronomical distance is the *light year*, the distance travelled by light in one year. It is equal to the speed of light, 186 000 miles or 300 000 kilometres per second, multiplied by the number of seconds in one year. Thus the light year is equivalent to nearly six million million miles or 9·5 million million kilometres.

$$1\ \textit{parsec} = 3{\cdot}26\ \textit{light years}$$

The nearest star to the sun is *Proxima*, the faint companion of the bright star Alpha Centauri, at a distance of 4·3 light years or 1·3 parsecs. Compare this with the sun's distance of about 8·3 light minutes. The figures for 61 Cygni and some well known first-magnitude stars are given below.

TABLE 7.1
DISTANCES OF SOME BRIGHT STARS

STAR	PARALLAX IN SECONDS	DISTANCE IN PARSECS	DISTANCE IN LIGHT YEARS
ALPHA CENTAURI	0·760	1·4	4·3
SIRIUS	0·379	2·6	8·6
61 CYGNI	0·298	3·4	11·1
VEGA	0·123	8·1	26·4
ARCTURUS	0·090	11·1	37·6
ALDEBARAN	0·048	20·8	67·9
ANTARES	0·019	52·6	172·0
BETELGEUSE	0·005	200·0	652·0

SIZES OF THE STARS

Because the stars are so far away that even the largest telescope in the world fails to reveal them as anything more than points of light, the finding of their diameters is a very difficult matter. However, with the aid of an instrument called an interferometer, astronomers have been able to make some direct measurements. Established laws of physics are also used in order to deduce the diameter of a star from the amount of energy it radiates, and fortunately the results obtained in this way closely agree with those determined by direct measurement. Therefore, the method of deduction has been extended to stars beyond the scope of the interferometer. Another method available is the study of the light variations of eclipsing binaries—two stars revolving around one another in such a manner that each star in turn blocks out light from the other as seen from the earth. Results from the various methods in use show that there are enormous differences in the sizes of stars, ranging from tiny *dwarfs* no bigger than the earth to fantastic *super giants* many hundreds of times bigger than the sun. For instance, if the sun was replaced by the well known star Antares it would extend beyond the orbit of Mars; other stars, not familiar by name, would even engulf Saturn.

COLOUR AND LUMINOSITY OF THE STARS

A casual examination of the night sky reveals stars of various colours: Vega has a bluish tint; Sirius is distinctly white; Capella is yellow; and Aldebaran has a reddish hue. If an iron poker is heated sufficiently it becomes "red hot"; if more heat is applied it becomes "white hot". Now, in the same way as the colour of the poker is related to its temperature, so the colour of a star is a guide to its temperature too. Blue stars and white stars are very hot, while red stars are comparatively cool, with yellow and orange stars lying somewhere in between.

When a stone is thrown into a pond it causes ripples on the surface of the water which appear to radiate outwards from the point of impact. The energy radiated by a star travels in a similar kind of waveform but with the speed of light, the distance between the crests being called the *wavelength*. This radiant energy actually consists of a variety of wavelengths and, apart from the small portion of *visible light*, cannot be detected by the human eye. Ordinary white light when passed through a glass prism is split up or dispersed into a *spectrum* which is composed of seven constituent colours in order of colours of the rainbow—red, orange, yellow, green, blue, indigo and violet. Different colours have different wavelengths. The shorter wavelengths lie at the violet end of the spectrum; beyond the violet the radiant energy consists of invisible and extremely short waves such as ultra-violet light, X-rays and gamma rays. Towards the red end of the

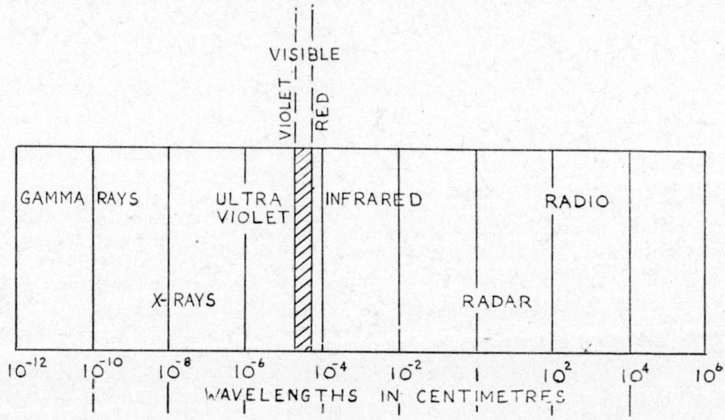

Fig. 7.2
Radiation in a Variety of Wavelengths.

spectrum, wavelengths become progressively longer as they merge into the invisible infra-red, radar and radio bands (fig. 7.2). An instrument called a *spectroscope* allows the light received from a luminous source to be analysed by splitting it up into an arrangement of lines known as *spectral lines*. If the source is a luminous solid like the filament of a glowing electric light bulb, the spectrum is *continuous* from the red through to the violet. If the source is a glowing gas, each chemical element of which the gas is composed will give an arrangement of *bright lines* in the spectrum by which it can be identified. For example, if salt (sodium) is sprinkled into a gas flame, two bright yellow lines characteristic of sodium will appear in the spectrum against a dark background (fig. 7.3). If the glowing lamp is then placed behind the sodium flame a continuous spectrum will be present, but in place of the two bright yellow lines there will be two dark lines in exactly the same position; the cooler sodium flame is now *absorbing* some of the energy being emitted by the brighter lamp behind it. In other words, the wavelength at which an element emits is exactly the same as that at which it absorbs. Thus the position of the lines in the spectrum allows different elements to be identified, just as human beings are by their finger prints.

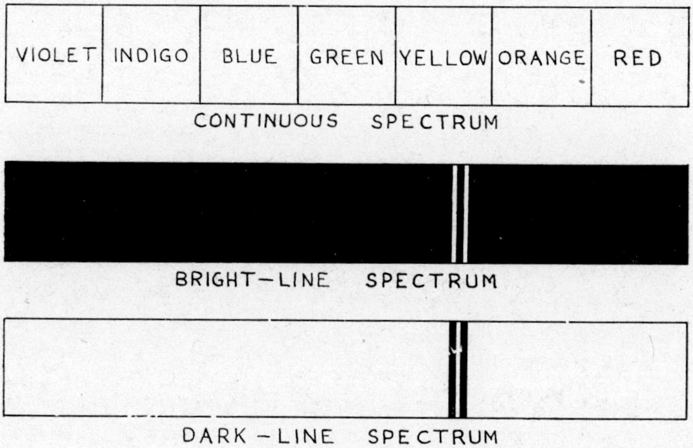

Fig. 7.3
Continuous, Bright Line, and Dark Line Spectra.

A star exhibits a dark line spectrum, and from the study of the spectral lines in starlight it is possible to determine the elements of which a star's atmosphere is composed. From the intensity of radiation in various wavelengths it is possible to calculate a star's temperature, and the results confirm that the colour of a star is definitely linked to its temperature. It has been found that the spectral lines of different elements are either very prominent or extremely weak depending upon a star's temperature: hot stars exhibit strong lines of hydrogen and helium, but as the temperature becomes lower the helium lines disappear and the hydrogen lines become weaker, while the lines of metallic elements get stronger. This has led to the grading of stars according to spectral type, some of which are given below as a guide.

TABLE 7.2
SPECTRAL TYPES—TEMPERATURES AND COLOURS

SPECTRAL TYPE	APPROX. SURFACE TEMPERATURE	COLOUR
O	40 000°	blue
B	20 000°	bluish-white
A	10 000°	white
F	7000°	yellowish-white
G	5500°	yellow
K	4000°	orange
M	3000°	red

Numerical subdivisions have been introduced to cover the transition from one type to another. The sun is classified as a G_2 star, the spectrum of which is characterised by strong calcium and metal lines. The dark lines in the sun's spectrum, first noted by Fraunhofer in 1814, are known as *Fraunhofer lines*.

In Chapter 2 we saw that the *apparent magnitude* of a star is a measure of its brightness as it is seen in the sky. In order to compare the real or intrinsic brightness of the stars, it is necessary to think of them all as being at the same distance; the distance chosen is 10 parsecs or 32·6 light years. The magnitude a star would have at the standard distance is called its *absolute magnitude*. The apparent magnitude of the sun is —26·7, but at the standard distance it would be only +4·8, *i.e.* just faintly visible to the naked eye. The absolute magnitudes of several well known stars are: Sirius 1·5; Canopus —4·4; Alpha Centauri 4·1; Arcturus —0·3; Vega 0·5; Capella —0·6; Betelgeuse —5·8; Antares —2·6; Deneb —4·8; and Regulus —0·7. Rigel, with an absolute magnitude of —6·4, is over 10 000 times more luminous than the sun

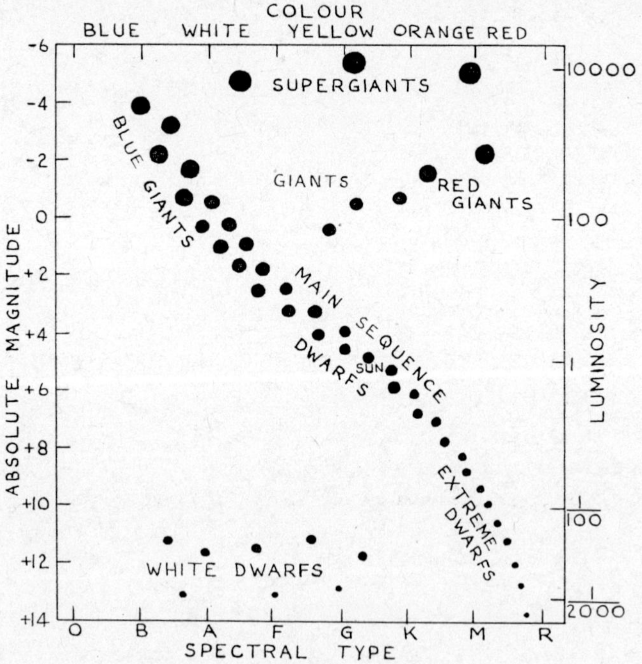

Fig. 7.4
Hertzsprung-Russell Diagram.

which means that it must have about 10 times the diameter of the sun. (For the purpose of comparing luminosity, the value for the sun is taken as 1.)

The relation between spectral type and absolute magnitude is best illustrated on a *Hertzsprung-Russell* diagram (fig. 7.4), in which absolute magnitude (luminosity) is plotted against spectral type (colour). When the stars for which data is available are so plotted it is found that the majority of them lie on a band running diagonally across the diagram. This is called the *main sequence.* The sun, a G_2 yellow star, lies on the main sequence about halfway along, showing that it is about 100 times cooler than the average blue star and about the same amount brighter than the average red star. On the main sequence, blue stars are large and hot (giants), while red stars are small and relatively cool (dwarfs). However, off the main sequence many exceptions are to be found: red supergiants such as Betelgeuse; blue supergiants like Rigel; and white dwarfs which are far less luminous than corresponding main sequence stars. Thus we find an enormous variety among the stars, a study of which leads us to the unavoidable conclusion that as a star our "very important sun" is really rather ordinary.

STARS OF SPECIAL INTEREST

Double stars; stars of variable brightness; pulsating stars; explosive stars—these are objects possessing unusual characteristics which make them worthy of special consideration.

DOUBLE STARS

Many stars which appear single to the naked eye are seen to be two stars when examined through a telescope. Although the existence of such stars had been noted in the early 17th century, the first systematic investigation of *double stars* was carried out by Sir William Herschel around 1780. Since there are so many stars to be seen, it is inevitable that many of them appear to form pairs. A large number of these appear double simply because the stars forming a pair are nearly in the same line of sight. In reality they are very far apart with no physical connection between them; such pairs are called *optical double stars.* Some stars, however, are truly double and consist of two close stars united by their mutual gravitational attraction. Like the earth-moon system, they revolve around each other about a common centre of gravity, thus supporting the claim that the law of gravitation acts beyond the solar system. Stars physically linked in this way are called *binaries: visual binaries* are those which can be separated into their com ponent stars by means of the telescope; *spectroscopic binaries* are those whose components are too close together to be separated visually but which have been found to be double by means of the spectroscope. When a source of light is approaching an observer the waves become crowded together and the wavelength is decreased. If the light is fed through a spectroscope, the effect is that the spectral lines are shifted slightly towards the violet end of the spectrum. On the other hand, the wavelength is increased when a source of light is moving away and the spectral shift is towards the red. A similar effect occurs in sound waves, evidenced by the change in the pitch of the siren signal as a train or a fire-engine rushes past, and is called *Doppler effect* (Johannes Doppler, 1842). When the stars of a binary system whose plane of revolution is tipped towards the earth are at their greatest distance apart, one is approaching us and the other is receding from us. Doppler effect causes the spectral lines of the approaching star to exhibit a *blue shift* and those of the receding star a *red shift*, resulting in a double spectrum. Later, when the stars come into line with the earth, the spectral lines close up, and when the stars are in reversed positions the spectral lines are displaced in the opposite directions. Thus a double star which cannot be separated in the telescope betrays its true nature in its spectrum. The periods of revolution of double stars range from a few hours for some spectroscopic binaries to many centuries for some visual binaries.

VARIABLE STARS

A *variable star* is one whose brightness does not remain constant with the passing of time. Such stars, of which more than 20 000 are known, may be divided into several main classes according to the nature of their variability: eclipsing binaries; pulsating stars; long-period and irregular variables; and explosive stars.

An *eclipsing binary* is not intrinsically variable; it appears to be variable simply because the plane of its revolution is inclined towards us so that as the stars revolve they regularly eclipse one another, either partially or totally. In early times, the Arabs were familiar with a strange star whose brightness varied and they called it *Al Ghul*, meaning *the ghoul* or *demon star*. We know it as Algol, or beta Persei. In 1872, John Goodricke, a young English astronomer, suggested that the fluctuations in Algol's brightness were caused by a bright star being partially eclipsed by a fainter companion, a theory which was shown to be correct by means of the spectroscope over a period of 100 years later. Algol normally has a magnitude of 2·3, but every 69 hours it fades in about five hours to magnitude 3·4, returning to normal after a further five hours. There is also a slight fluctuation in brightness halfway through the 69 hour period, but this can only be detected with special equipment. The large fall and rise occurs as the fainter companion is eclipsing the brighter star; the minor fluctuation as the brighter star is eclipsing the companion (fig. 7.5).

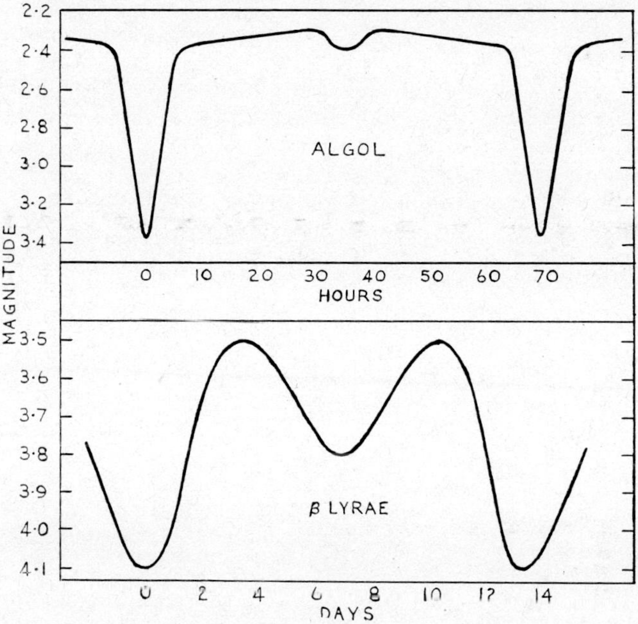

Fig. 7.5
Light Curves of Algol and Beta Lyrae.

Owing to a wide variety in the sizes, magnitudes and separations of the stars in binary systems, and the inclination of their orbits, the light curves of many eclipsing-type variables differ considerably from that of Algol.

As we have seen, the stars of a binary system are not in themselves variable, the fluctuations in brightness being due to the fact that each star in turn blocks out light from the other. On the other hand, many stars are *intrinsically variable* because they alternately expand and contract, with a consequent rise and fall in their energy output. Some of these intrinsic variables actually pulsate with clockwork regularity, and the first star of this type to be discovered was delta Cephei. From its normal apparent magnitude of 3·6, delta Cephei dims to a magnitude of 4·3 and then returns to 3·6 in just over five days.

Its light curve (fig. 7.6) is characterised by a fairly rapid rise to maximum brightness followed by a slower fall to minimum brightness. Other variable stars whose light curves resemble that of delta Cephei are known as *cepheid variables*. The periods of different *cepheids* range from as little as one day to nearly two months, but in each case the period—time between peaks of maximum brightness—is strictly rhythmic. Another class of periodic variables are the *RR Lyrae stars*, or *cluster variables*, which have periods of less than one day and in some cases less than two hours. In the early part of this century Miss Leavitt, an American astronomer, noted a relationship between the periods and absolute magnitudes of the cepheids in the Magellanic Clouds—two star clouds in the southern celestial hemisphere—

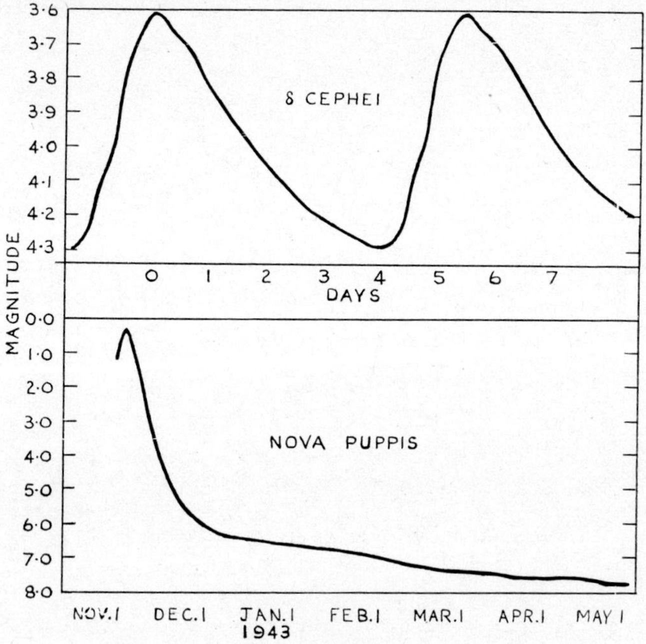

Fig. 7.6
Light Curves of Delta Cephei and Nova Puppis.

such that the period is greater as the stars are more luminous, *i.e.* smallest for the shortest periods and greatest for the longest periods. This led to the formulation of the *period-luminosity law*, which links the periods and luminosities of all cepheids no matter where they occur in the universe.

If the absolute and apparent magnitudes of a star are known, it is possible to calculate its distance: in the case of a cepheid the apparent magnitude is observed and the absolute magnitude is obtained by the period-luminosity law from the observed period of the star. Thus the cepheids provide astronomers with the means of finding the distances of remote star groups and galaxies lying beyond the range of parallax measurement. It was thought that the original period-luminosity law applied to all cepheids, but it was found later that some cepheids follow another form of the law. Those which resemble the prototype are the *Type* 1, or *classical cepheids:* the others, including the RR Lyrae stars, form *Type* 2. The application of the same law to all cepheids led to considerable errors in the calculated distances of many objects, a matter which has only been put right in recent years.

Many red giants and supergiants form two classes of stars called *long-period variables* and *irregular variables*. The best known of the long-period type is Mira Ceti, meaning "the Wonderful star in Cetus". It fluctuates irregularly in brightness from 2nd and 5th magnitude at maximum to 8th and 10th magnitude at minimum, which means that Mira varies from a naked-eye star to a telescopic object; its period averages 330 days. Long-period variables have periods greater than a few months, though mostly around 300 days with a range in brightness of about five magnitudes. A notable exception is Chi Cygni, which can range over 10 magnitudes, but for most of its 400 day cycle it remains below naked-eye visibility. A good example of an irregular variable is the red star Betelgeuse in the constellation of Orion, which varies in magnitude from 0 to 1·2 in a period which can be as much as six years. Measurements with the interferometer show clearly that Betelgeuse is pulsating, its diameter varying from around 300 to 400 times that of the sun.

A *nova*, or "new star", is one which explodes violently, acquiring a temporary superior brilliance—an increase of 10 magnitudes, or 10 000 times the former luminosity, is not unusual—before fading back into comparative obscurity. Nova Puppis (1942) rose so rapidly that it acquired 1st magnitude brightness before it was noticed. It increased further to magnitude 0·3 and faded back to 6th magnitude in the space of about a month, taking several more months to return to somewhere near its normal 8th magnitude.

Not all novae follow the same pattern: Nova Herculis was discovered on December 13, 1934, when just visible to the naked eye. By December 22 it had brightened to 1st magnitude, after which it faded slowly but remained visible to the naked eye until April of the following year when it quickly fell to 13th magnitude. It recovered to near naked-eye visibility before the final decline set in at the beginning of 1936. More recently, Nova Delphinus (1967) was discovered at just above naked-eye level. It attained 4th magnitude quickly and remained at maximum for about a month with only minor fluctuations in brightness, followed by a gradual decline. It is believed that this nova was a 12th magnitude star before its eruption.

A *supernova* is a star which explodes with much greater violence than an ordinary nova, perhaps growing temporarily in luminosity to 100 million times its original intensity. Tycho Brahe observed a remarkable supernova in 1572 which grew so bright that it rivalled Venus; Kepler also observed one in 1604. In 1054, Chinese astronomers recorded the appearance of a supernova in Taurus which attained the brightness of Jupiter and remained visible for two years. In this region of the sky there is an expanding gas cloud known as the Crab Nebula, so named because of its crab-like appearance. It does give the impression of a vast explosion having taken place, the star which met catastrophe probably being one of the faint stars near the centre of the nebula.

THE GALAXY

The sun and its planetary system is a minor part of a massive stellar family called the *Galaxy*, which consists of over 100 000 million stars and large quantities of dust and gas, held together by gravitation.

THE MILKY WAY

The stars of the Galaxy are distributed in the form of a flattened disc about 80 000 light years in diameter, with a spheroidal central nucleus about 10 000 light years in thickness—in shape, the Galaxy is not unlike two cymbals placed face to face (fig. 7.7). Spiral arms of stars and gas unfold from the nucleus, and it is within one of these arms that the sun is situated about 27 000 light years from the *galactic centre*. In the region of the sun the thickness of the Galaxy is about 2000 light years. The whole system rotates slowly about the galactic centre, the sun's period being several hundred million years.

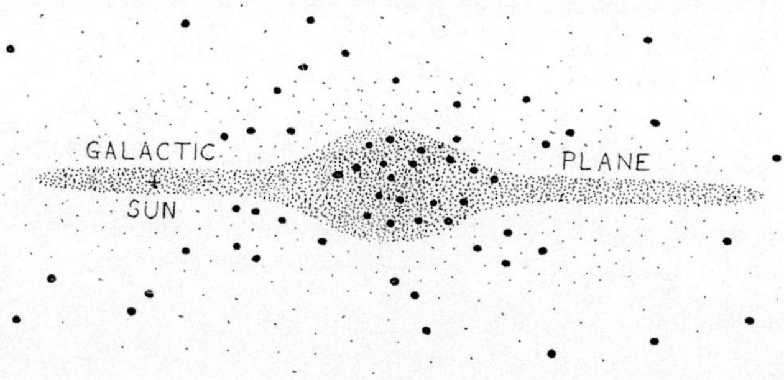

Fig. 7.7
Structure of the Galaxy and Distribution of Globular Clusters.

From the earth, an observer looking along the plane of the Galaxy will see a much greater density of stars than he would if he looked into a direction at right angles to it. The majority of the stars are too far away to be seen individually, but along the galactic plane the light of millions of faint stars combines to form a pale luminous band called the *Milky Way* which, varying in brightness and width, stretches across the night sky from horizon to horizon. The Milky Way follows roughly a great circle around the celestial sphere, inclined at an angle of 62° to the celestial equator. From the northern constellation of Cassiopeia, it passes through Cepheus and into Cygnus (Northern Cross) where it splits up into two distinct branches. One branch runs through Aquila and Sagittarius, and the other through Ophiuchus and Scorpius. In the region of Crux (Southern Cross) the two branches merge and then continues as a single band through Carina, Puppis, Canis Major, Taurus, Auriga, Perseus and back to Cassiopeia.

GALACTIC NEBULAE

The space separating the stars from one another is by no means empty of matter. Within the plane of the Galaxy, and in particular in the spiral arms, there are vast clouds of dust and gas known as *nebulae*. They do not emit light of their own, although in many cases they are caused to glow by bright stars in their vicinity. Originally, the word nebulae was applied to objects, other than comets, which appeared as hazy clouds when seen through the telescope. They were catalogued by a French astronomer, Messier, who was anxious to ensure that they should not be confused with comets. Nowadays, these objects are included in the *New General Catalogue*, but many are still popularly known by their Messier numbers. Thus the Crab Nebula, known as M1 because it is the first object in the Messier list, is also NGC 1952. Many of the objects in the original Messier list have since been recognised as star clusters or galaxies of stars lying beyond our own system. Galactic nebulae are of two main types: bright and dark.

Bright nebulae are dependent for their luminescence upon the radiation of nearby stars which cause the nebular gases to glow with a pale greenish light. Some bright nebulae are diffuse clouds such as the Great Nebula in Orion, situated near the middle star in the Sword of Orion at a distance of over 1000 light years from us. In the telescope it is visible as two greenish clouds of unequal size, in which are embedded a number of stars. *Planetary nebulae*, also classed as bright, usually assume the appearance of a disc or ring of bluish-green gas surrounding a hot blue star whose radiation is responsible for the illumination of the gas. An excellent example is the Ring Nebula in Lyra. The name "planetary" was given to this type of nebulae in the early days of telescopic observation because their appearance resembled that of the outer planets Uranus and Neptune.

Dark galactic nebulae are normally invisible because there are no nearby stars to illuminate them, and they can only be detected where they appear against a brighter background. They are responsible for the Great Rift in the Milky Way and other dark patches such as the Coal Sack in the Southern Cross. A spectacular case is that of the Horse-Head nebula in Orion.

STAR CLUSTERS

In the early 18th century, Edmond Halley noted that some bright stars were as much as $\frac{1}{2}^\circ$ away from the positions given in Ptolemy's Almagest. This was the first indication that the stars are not really "fixed" but moving in space relative to each other. The component of a star's space motion in the line of sight, either towards us or away from us is called its *radial motion*. The change in a star's position on the celestial sphere due to its movement at right angles to the line of sight is called its *proper motion*. In most cases the stars comprising constellations are at greatly different distances from us, the familiar patterns we see being peculiar to our position in space, but their proper motions carry them in various directions relative to each other so that in tens of thousands of years time the constellations will not look quite the same as they do now. On the other hand, *clusters* are collections of stars less widely separated from each other, the members of which are related by virtue of possessing similar proper motions. This means that clusters will maintain their identities for much longer than the larger constellations. There are two principal types: *galactic clusters* and *globular clusters*.

Galactic clusters are so named because they are mostly found near the galactic plane; within 20 000 light years from the sun several hundred are recognised. They are also known as *open clusters* because the stars in them are more widely spaced compared with the more compact globular type. The nearer clusters having larger proper motions are sometimes called *moving clusters*. The best known of the open type is the Pleiades, or "Seven Sisters" in Taurus, which is situated about 500 light years from the sun. Persons with normal sight can see about six or seven stars with the unaided eye, but a pair of binoculars or a small telescope reveals a beautiful concentration of stars. The largest and nearest cluster is the Hyades, the familiar V-shaped group of stars near Aldebaran, also in Taurus. It covers a region of the sky about 25° in diameter, though most of its 150 stars are very faint; Aldebaran itself is not a member of the cluster. Other notable open clusters are: the Praesepe or "Beehive" cluster in Cancer, and the double cluster in Perseus. Fig. 7.8 shows that five of the

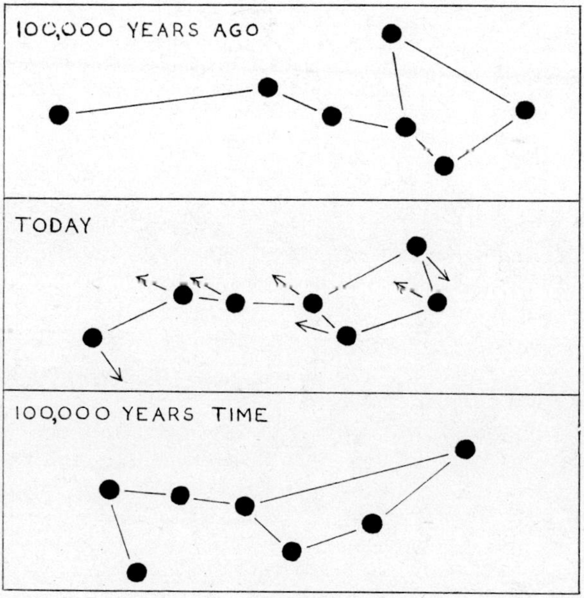

Fig. 7.8
Proper Motion of the Stars of The Plough.

bright stars of The Plough form a moving cluster, but because the common proper motion is not shared by the end stars Dubhe and Alkaid the shape of the constellation will not always be the same as it is now.

Globular clusters consist of many thousands of stars concentrated into a relatively small spherical volume of space. Over a hundred or so of such clusters are recognised, the majority of which are not visible to the naked eye. Those located in the southern constellations Centaurus and Tucana, and in the Northern constellation Hercules (M13), are the brightest and may be seen as faint nebulous patches without optical aid. Globular clusters are very distant, the nearest being over 20 000 light years away. They are unevenly distributed throughout the Galaxy, noticeably absent from the galactic plane but strongly concentrated on either side of it, being particularly numerous in Sagittarius where our line of sight is directed towards the galactic centre. In the vicinity of the sun, distances between neighbouring stars are about several light years; in condensed open clusters about one light year; and in globular clusters as little as several light weeks or even light days near their centres.

BEYOND THE GALAXY

Our own Galaxy is but one among millions of similar "island universes" spread through the depths of space apparently without limit; for whenever the frontiers of astronomy are extended by the introduction of larger telescopes, yet more galaxies are discovered.

EXTRAGALACTIC OBJECTS

Galaxies fall into either of three main classes according to their telescopic appearance: *elliptical galaxies; spiral galaxies;* and *irregular galaxies.* Elliptical galaxies are smooth in appearance with fairly bright centres and fainter edges, their outlines ranging in shape from circular to pointed ellipses. They are designated by the letter E followed by a number which indicates the amount of their ellipticity, the circular ones being called E_0 and the most elongated E_7. Spiral galaxies, as their name implies, have spiral arms of stars, dust and gas. Normal spirals possess a central nucleus, from which the arms emerge and coil around the centre, rather like a "Catherine Wheel". Three principal types are recognised: Sa galaxies having large centres with tightly wound arms; Sb galaxies having smaller centres and more open arms; Sc galaxies having small centres and widely opened arms. Barred spirals have elongated central regions with spiral arms extending from each end of the bars. Like the spiral galaxies, they are designated according to the amount of their compactness; SBa, SBb and SBc. Galaxies having no regular shape are classified as irregular and designated by the letter I. The broad classification described above has been added to in the light of modern investigations of the stellar populations of galaxies.

The nearest and brightest of the external galaxies are the Large and Small Magellanic Clouds, which are plainly visible with the unaided eye to observers in the southern hemisphere. They were referred to in the log of the intrepid Portuguese navigator Ferdinand Magellan, after whom they were named. The Large Cloud, situated in the constellation Dorado, has a diameter of about 30 000 light years; the Small Cloud in Tucana, is about 20 000 light years across. The Magellanic Clouds were mentioned earlier in connection with cepheid variables and the period-luminosity law, from which they have each been found to be about 170 000 light years distant. Both Clouds are irregular galaxies and by astronomical standards they are close enough to be regarded as companions of our Galaxy. The most distant object in the sky visible with the naked eye is NGC 224, the Spiral Galaxy in Andromeda, also known as M31 or the Great Nebula in Andromeda. On a dark, clear night it may be seen as a small, faint patch of light situated a little to the north of the star beta Andromeda, but with the world's largest telescopes its spiral arms can be resolved into stars. Fortunately, some of these are cepheids, from which its distance has been calculated to be nearly two million light years. It also contains clusters, and several temporary stars have been observed, including a supernova. M31, an Sb spiral galaxy, is about 120 000 light years across, and apart from its smaller size our Galaxy is thought to be very similar to it. Within the photographic range of the world's largest telescopes there are thousands of millions of galaxies scattered in all directions through space except for a *zone of obscuration* in the galactic plane. At first, when astronomers knew of the existence of comparatively few galaxies, it was assumed that they were distributed at random, but it is now known that they tend to be grouped in clusters. Our Galaxy belongs to a small local group comprising the Magellanic Clouds, M31 and its companion M32, the Triangulum Spiral M33, and about a dozen smaller galaxies, all situated within a sphere of space about two million light years in diameter. American astronomers recently discovered two large galaxies about twice as distant as the Andromeda Galaxy. They are almost obscured by huge dust clouds in the Milky Way, and they have been named Maffei 1 and 2 after an Italian astronomer who first observed them in 1968 but did not know what they were. Next beyond the local group, at a distance of about 20 million light years, lies the Virgo cluster, containing over a thousand closely packed galaxies in the direction of the constellation Virgo. Some constellations, such as Ursa Major, contain more than one cluster.

THE EXPANDING UNIVERSE

With the exception of a few of our nearest neighbours, the external galaxies exhibit a red shift in their spectra. The generally accepted reason for this is that they are moving away from us at high speeds, the speed of recession being greater the further away they are. According to one theory the whole of the matter of the universe was congregated in one place several thousand million years ago. The mass of matter became unstable and exploded, throwing material outwards in all directions which later condensed into galaxies and individual stars (and planets). This is known as the *Big Bang theory*. Another version is that the rate of expansion of the universe is slowing down so that at some time in the distant future the galaxies will begin to move back towards each other, rather like stones thrown up into the air returning to the ground under the influence of gravity. Eventually, the matter of the universe will come together and explode again to give birth to a new universe. This life and death cycle will go on repeating itself for all eternity according to the *Oscillating Universe theory*. The great rival of the Big Bang is the *Steady State theory*, which holds that the universe is infinitely large and infinitely old. It accepts that the universe is expanding but states that new matter is being continuously created so that as older, more distant galaxies move away, new galaxies are formed to fill the space left. Thus the universe appears much the same now as it always did and always will appear more or less the same. This theory conflicts with the long held belief that matter can neither be created nor destroyed. If the frontier of the receding galaxies could be detected it would lend support to the Big Bang. On the other hand, if the outermost galaxies achieve the speed of light they would become invisible to us, and it could then be argued that the new matter created according to the opposing theory would keep the visible universe in a steady state. Astronomers are striving continuously to find evidence that will support one theory or the other, but at the present time no conclusion is possible.

THE RADIO SKY

Most of the energy emitted by distant stars and galaxies is impeded by the earth's atmosphere, and so far we have concerned ourselves only with the small "window" through which the visible light reaches us. However, there is another window in the longer wavelengths beyond the infrared which allows energy in the form of radio waves to penetrate through the atmosphere to the earth; this is the concern of radio astronomers. Radio astronomy had its beginnings in 1931 when Karl Jansky, an American, discovered a weak hissing signal in his radio equipment having a period of one sidereal day. The signal strength was greatest each time the earth's rotation carried the radio aerials around until they were facing the constellation Sagittarius, the general direction of the galactic centre. In 1937, Grote Reber, also an American, constructed a dish-shaped aerial 31 ft (9·5 m) in diameter specially for the purpose of receiving these signals from outer space. He confirmed the strong signals from Sagittarius and also found a number of other localised sources. During the Second World War, J. S. Hey traced interference in the vital British radar system to the sun; the first time a radio source was linked directly with a visible object. Shortly after the war, the science of radio astronomy developed very rapidly on both sides of the Atlantic and later in other countries, notably Australia. Numbered among the world's largest radio telescopes are: the 329-ft (100-m) fully steerable dish at the Max Planck Institute, Western Germany; the 250-ft (76·2-m) fully steerable dish at Jodrell Bank, England; the 300-ft (91·4-m) partially steerable dish at Greenbank, USA; the 150-ft (46·7-m) dish at Algonquin, Canada; the 210-ft (64-m) dish at Parkes, Australia; and the fixed 1000-ft (305-m) dish built into a natural crater at Arecibo in Puerto Rico. Radio telescopes are often linked together in such a way that they give results equivalent to that of a dish having a diameter of the length of the line joining them; a method known as *baseline interferometry*. Very long baseline interferometry, VLB, pioneered by the Canadians across their Continent from Algonquin in the east to Penticton in the west, now connects radio telescopes across the world, the longest link being from Algonquin over the Pacific to Parkes.

Radio energy emanates mainly from hydrogen and electrified particles travelling at high speed through magnetic fields. The signals, often extremely weak, are received by radio telescopes, amplified and then fed to recording apparatus. One of the first tasks of the radio astronomers was to map the radio sources in the Galaxy, and naturally enough they were found to be plentiful in the galactic plane. A radio map consists of lines, called *isophotes*, joining points having the same intensity of radiation. Intense sources, such as Taurus A in the Crab Nebula, show up as tightly packed closed curves. Of course, radio and optical astronomers work in close harmony. For example, in 1951, Cambridge radio astronomers accurately fixed the position of the second strongest radio source, Cygnus A. The information was passed to observers at the Palomar telescope and a photograph revealed what appeared at first to be two galaxies in collision; though it is now thought to be a single galaxy exploding. The radio object consists of two radio emitting clouds, together some 20 times larger than the visible object. It is estimated to be 500 million light years away, but as a radio source it is far stronger than the Crab Nebula which is only 4000 light years distant—in fact it would take 10 million "radio Crabs" to make one Cygnus A. It is quite significant that many of the strongest radio sources are located in galaxies undergoing some violent change, another example being M87 in Virgo which has ejected a huge jet of material.

In 1960, Jodrell Bank astronomers detected intense radio sources much smaller than the radio galaxies. American astronomers identified one of these, 3C48 (the 48th object in the 3rd Cambridge catalogue), with a blue star. This was the first of many starlike radio sources to be discovered; given the name of *quasi-stellar objects*, or *quasars*. At first they were thought to be in our own galaxy but events later proved them to be extragalactic. In 1962, radio astronomers at Parkes were able to pin-point the exact position of 3C273 when it was occulted by the moon. Palomar astronomers found in the position indicated a blue star with a jet protruding from it, the star and the jet corresponding to a double radio source found at Parkes. Examination of the spectrum of this quasar revealed a fantastic red shift, from which it was estimated to be 1500 million light years distant with a speed of recession far in excess of that of any other known object. Further study of the spectra of other known quasars showed that they too were red shifted, some even more than 3C273. Thus the quasars were recognised for the fantastic objects they are: the most distant objects in the universe, possessing such enormous concentration of energy that one of them would make 10 million million suns. There is evidence to suggest that quasars show a tendency to cluster, notably around a distance of about 8000 million light years with a speed of recession in the order of 80 per cent of the speed of light. It is also believed that many blue stars previously thought to be near the edge of our Galaxy are also quasars, although they are not detectable radio sources. On the other hand, many radio sources not related to visible objects have been identified as quasars. Some astronomers believe that quasars represent an early stage in the evolution of galaxies, but at present there is no general agreement on the nature of these remarkable objects.

In 1968, Cambridge astronomers discovered the first of the *pulsating radio sources;* stars that radiate great bursts of radio energy with the regularity of a ticking clock. Australian scientists working with the Mills Cross, an extensive array of aerials in the form of a cross, located many more of these unusual objects which have become known as *pulsars.* One of these, Vela X, lies in a region of the sky occupied by an exploded star. Later in 1968, astronomers at Greenbank and Arecibo discovered a pulsar in the Crab Nebula, thus lending support to the theory that there is a connection between pulsars and supernovae. It was in the Crab Nebula that the first visible sighting of a pulsar was made in 1969 by astronomers at the Steward Observatory in Arizona; the star in question is thought to be the remains of the one which exploded and caused the nebula. The pulsations are very rapid, several times a second in some cases, and this had led to the belief that pulsars are small, extremely dense objects.

REVIEW QUESTIONS

1. State how the following can be related indirectly to the sun:
   (a) The ability of a glider to stay in the air.
   (b) The paper on which this book is printed.
   (c) The leather from which shoes are made.

2. Why do sunspots appear to move towards the sun's western limb? When will the next sunspot maximum occur?

3. What is parallax of a star? Why should the parallax method of determining stellar distances be limited to about 300 light years?

4. Name two methods by which it can be proved that the earth revolves around the sun.

5. If the sun stays on the main sequence, what kind of star will it eventually become? In what respects does the sun differ from the star Betelgeuse?

6. Distinguish between absolute magnitude and apparent magnitude.

$$M = m + 5 - 5 \log d$$

Where M is absolute magnitude, m is apparent magnitude and d is the distance of a star in parsecs.

If the star Vega is at a distance of 8·1 parsecs and its apparent magnitude is +0·4, calculate its absolute magnitude.

7. Knowing that Vega's distance is 8·1 parsecs, calculate its parallax and distance in light years.

8. The visual luminosity of a star relative to the sun is given by:

$$\log l = 0{\cdot}4\,(4{\cdot}8 - M)$$

Where l is the star's luminosity and M its absolute magnitude. If the absolute magnitude of Capella is —0·6, how much more luminous is it than the sun?

9. From the formula given in question 6, show that:

$$\log d = \frac{m - M + 5}{5}$$

A cepheid variable has an apparent magnitude of +10 and an absolute magnitude of —3·5. Calculate its distance.

10. Distinguish between an optical double star and a visual binary. Are all binaries of the eclipsing type?

11. Enumerate the different classes of variable stars and name one of each type.

12. Which of the following are galactic and extragalactic: (a) Crab Nebula, (b) Orion Nebula, (c) pulsars, (d) quasars, (e) Andromeda Nebula, (f) Magellanic Clouds, (g) Hercules cluster?

PROJECTS AND RESEARCH

1. THE SUN

Without the light and heat of the sun, no form of life would have evolved on our planet and the earth would be a dead, frozen world. Using the information given below as possible lines of research, draw up an account headed "The Sun and its Influence".

(a) The sun is a star. It is the dominant member of the solar system but a minor member of the Galaxy.
(b) Aurorae, radio-blackouts and variations in the earth's magnetic field are caused by the sun.
(c) Sources of terrestrial energy like timber, coal, oil and natural gas can be related to the sun.
(d) The sun causes the water cycle; invisible water vapour produced by the process of evaporation rises from the earth's surface, cools and condenses into clouds from which snow and rain fall, thus returning the water back to the earth. Turbines driven by water power are used to generate electricity.
(e) The sun regulates our daily lives.
(f) The heat of our bodies, gained from the food we eat and the clothes we wear, can be traced back to the sun.
(g) Solar batteries are used as a source of power in artificial satellites and spacecraft.

This project could be combined with Project 3, Chapter 3.

2. OBSERVING SUNSPOTS

This is a rather pleasant pastime and one which provides an opportunity to verify some of the statements made earlier. However, it must be clearly understood at the outset that looking at the sun through any form of optical aid must not be attempted; it is extremely dangerous and can result in permanent blindness for anyone foolish enough to do so. Sunspots occasionally grow large enough to be seen with the naked eye, especially at times of maximum solar activity, but even so observations should only be made using a dark filter, a piece of smoked glass or several layers of developed photographic film. Where a small telescope is being used the method is to direct the instrument towards the general direction of the sun, being very careful not to look through it. Hold a sheet of white card, or white paper clipped to a smooth baseboard, away from the eyepiece and position it so that the image of the sun is projected on to the card (fig. 7.9). Adjust the eyepiece outwards from the normal setting until the sun's disc and any spots upon it are brought sharply into focus. Varying the

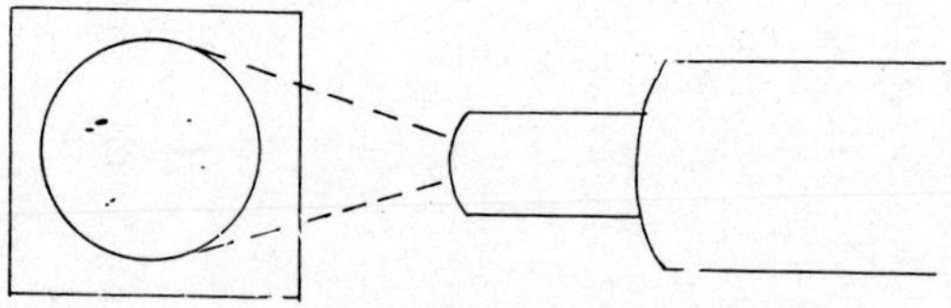

Fig. 7.9
Method of Projecting the Sun's Image.

distance of the card will allow you to decide the size of the image to work with—4 in or 10 cm diameter more or less—but having done this stick to the same image size for all observations. It is not envisaged that the reader has anything so grand as an equatorial, clock-driven telescope so the image will tend to drift as the sun performs its westward diurnal motion across the sky. This may be found inconvenient at first when it comes to drawing in the positions of the spots but with practice the problem will be overcome. It is also desirable, though not essential, that observations be made at about the same time each day because the attitude of the sun in the sky changes. Successive observations will show that the spots are carried around by the sun's rotation towards the western limb, the paths they follow representing parallels of solar latitude. Since the sun's axis of rotation may be tipped towards or away from us by up to $7\frac{1}{4}°$ and inclined to the east or west by up to $26\frac{1}{2}°$, the paths followed by the spots appear to change during the course of the year (fig. 7.10). Small black pores are quite common, as are small circular spots having a central umbra

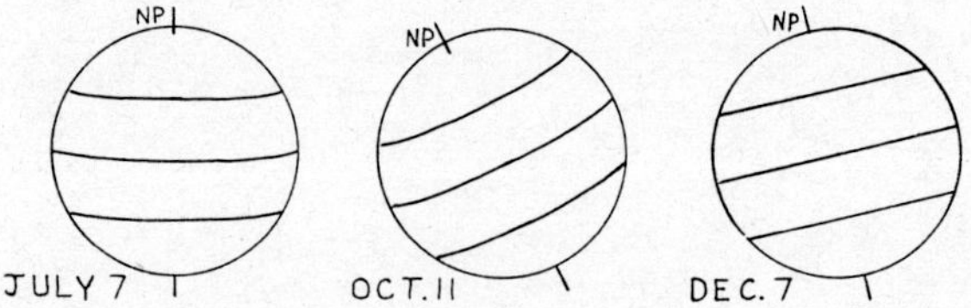

Fig. 7.10
Apparent Change in the Paths of Sunspots through the Year.

surrounded by a well-defined penumbra. It may be noticed that near the limb the spots appear oval-shaped with the darker umbra on the side of the spot away from the limb. This is perfectly normal because sunspots are depressions in the photosphere, their apparent distortion being an effect of changing perspective due to the curvature of the sun. It usually takes a spot about two weeks to traverse from the eastern limb to the western limb, provided it does not decay in the interval. Unusually large spot groups are of great interest because day to day changes in their appearance are often very noticeable. They may survive one or more rotations of the sun, disappearing over the western limb to reappear over the eastern limb about a fortnight later. Occasionally, disturbed areas without spots may be seen near the limb; these are called *faculae*. All observations should be timed, dated and accompanied by brief notes on the general appearance of spots seen.

3. OBSERVING AURORAE

Although aurorae occur in the earth's upper atmosphere, they originate from the streams of electrically charged particles emanating from the sun. These particles are channelled by the earth's lines of magnetic force towards the north and south magnetic poles, situated respectively in the Hudson Bay area of Canada and South Victoria Land in Antarctica. Thus aurorae are more common in higher latitudes; in the polar regions they are a regular feature of the night sky during the long winter months. In lower latitudes they are less frequent, spectacular displays being more likely around sunspot maximum when flares occur in greater numbers. The "northern dawn" is visible from the northern USA, and Canada, and also from Great Britain where observers in Scotland and northern England are more favourably placed than those in the south. Unfortunately, aurorae are denied to observers in low northern latitudes, and in the southern hemisphere their great beauty is wasted over the vast expanse of the Southern Ocean. Observations are best made on clear, moonless nights from locations well away from town lights. If a check is being kept on sunspots, there is an excellent chance of a display shortly after a large spot group is seen near the sun's central meridian. Telescopes and binoculars are of no advantage because they not only restrict the field of view but also reduce the light. From the middle-latitudes, an aurora most frequently appears as a pale glow along the northern horizon, but patches of light and vertical rays are also quite common. Sometimes the rays tend to move fairly rapidly, often converging to form a corona near the zenith. Other forms include arcs and bands extending from east to west across the sky, perhaps accompanied by vertical rays which may give the effect of a waving cosmic drapery. There is also a variety of colours in auroral light; grey-white, yellow-green, red and blue are all fairly common. Photography is possible, if rather chancy, and is more a case of "nothing ventured—nothing gained". The length of the exposure, which should be as long as possible (about 10 seconds), will obviously depend on whether the auroral form is stationary or moving.

4. MAPPING THE PLEIADES

Of the open clusters, the Pleiades in the constellation Taurus merits special attention because it is conspicuous to observers in both hemispheres. This compact group of stars has already figured in Projects 1 and 5 in Chapter 2 so locating it should present no problem, especially on a clear night between December and March, when it is best placed for observation.

TABLE 7.3
THE PLEIADES

STAR	MAG.	RIGHT ASCENSION	DECLINATION
PLEIONE	5·2	3 h 47·5 m	+ 24° 03′
ATLAS	4·0	3 h 47·4 m	+ 23° 58′
ALCYONE	3·2	3 h 45·7 m	+ 24° 01′
MEROPE	4·4	3 h 44·5 m	+ 23° 51′
ELECTRA	4·2	3 h 43·1 m	+ 24° 01′
CELAENO	5·6	3 h 43·0 m	+ 24° 11′
TAYGETA	4·7	3 h 43·4 m	+ 24° 22′
MAIA	4·4	3 h 44·0 m	+ 24° 16′
ASTROPE	6·2	3 h 44·1 m	+ 24° 28′

It is useful to construct a map on which the stars listed in Table 7.3 are plotted before any observations are made. Vertical lines of Right Ascension are spaced 1 in or 20 mm apart for each minute of time, from 3 h 40 m (right edge) to 3 h 50 m (left edge). Horizontal lines of declination are spaced 1 in or 20 mm apart for each 10′ of arc, from + 23° 30′ (lower edge) to + 24° 40′ (upper edge). A slight distortion in the shape of the cluster due to using a convenient scale rather than an exact one should not detract from the value of the exercise. Having plotted the positions of the stars on the map, observe the Pleiades with the unaided eye and see if you can locate each of the nine stars; if you can then you have exceptionally good eyesight. Most people detect only seven stars which is why the group is popularly known as the Seven Sisters. Many see only six stars, the elusive ones being Pleione, Celaeno and Astrope. The cluster actually contains well over 200 stars, and with the aid of a telescope or decent binoculars at least 50 of these could be added to the map.

5. STAR CLUSTERS AND NEBULAE

Examine the following list of clusters and nebulae and decide which of them are visible from your latitude. The position of each object is indicated on the star maps, but the advice given in Project 2, Chapter 2 may be found useful in determining when it is most suitably placed for observation.

NGC 104, 47 Tucanae. 0 h 23 m, — 72° 15′. (Map 10C and D).

The grandest of the globular clusters. Visible to the naked eye as a 5th magnitude star. Appears as a hazy patch of light in binoculars. Individual stars in a 3-in telescope.

NGC 224, M31, Andromedae. 0 h 41 m, + 41° 06′. (Map 4).

The Great Nebula in Andromeda. Visible as a hazy spot to the naked eye. Fuzzy oval shape in small instruments.

The Nubecular Minor. 0 h 51 m, — 73° 48′. (Map 10C).

The Small Magellanic Cloud in Tucana. Naked-eye object within which rich fields can be seen with binoculars and small telescopes.

NGC 869 and 884, Persei. 2 h 20 m, + 57° 01′. (Map 4).

The Double Cluster in Perseus. Visible to the naked eye, but magnificent in small instruments on a dark night in December.

NGC 1039, M34, Persei. 2 h 40 m, + 42° 37′. (Map 4).

A fine open cluster, just visible to the naked eye.

The Pleiades. 3 h 46 m, + 24° 04′. (Maps 3 and 4).

A magnificent cluster in Taurus. Shaped like a small saucepan. Best seen in low-powered instruments.

The Hyades. 4 h 34 m, + 16° 28′. (Maps 3 and 4).

V-shaped cluster in Taurus. Easily seen with naked eye but its stars are too widely separated to be imposing.

The Nubecular Major. 5 h 25 m, — 69° 20′. (Map 10C).

The Large Magellanic Cloud in Dorado. Naked-eye object with rich star fields for small instruments. The Large and Small Magellanic Clouds are rather like detached portions of the Milky Way.

NGC 1912, M38, Aurigae. 5 h 26 m, + 35° 48′. (Map 3).

A fine loose cluster in the form of a crucifix. Easily picked up with binoculars.

NGC 1976, M42, Orionis. 5 h 34 m, — 5° 24′. (Map 3).

The Great Nebula in Orion. A starlike object to the naked eye. Seen as a greenish cloud through small instruments. One of its central stars, θ Orionis, can be resolved into four components with a larger telescope.

NGC 2099, M37, Aurigae. 5 h 50 m, + 32° 33′. (Map 3).

A fine open cluster. Reddish star near its centre.

NGC 2168, M35, Geminorum. 6 h 07 m, + 24° 21′. (Map 3).

A large open cluster of faint stars, seen well with binoculars. May be seen with the naked eye.

NGC 2287, M41, Canis Majoris. 6 h 45 m, — 20° 42′. (Maps 3 and 9).

An open cluster which is easily visible with the naked eye. Reddish star near its centre.

NGC 2437, M46, Puppis. 7 h 41 m, — 14° 44′. (Map 3).

A fine cluster of faint stars.

NGC 2632, M44, Cancri. 8 h 38 m, + 20° 06′. (Map 5).

Praesepe, or "Beehive", cluster. May be just visible with the naked eye, but best seen through binoculars.

NGC 3372, Carinae. 10 h 44 m, — 59° 31′. (Map 10B).

The Keyhole Nebula. Fainter than the Orion Nebula, but may be located with good binoculars. Bright and dark nebula around Eta Carinae. Many faint stars.

NGC 3532, Carinae. 11 h 05 m, — 58° 30′. (Map 10B).

A magnificent cluster of stars. Best seen in telescopes of aperture greater than 3 in. Can be seen with binoculars.

NGC 3766, Centauri. 11 h 35 m, — 61° 26′. (Map 10A).

A fine cluster, well seen with binoculars and small telescopes.

NGC 4755, Crucis. 12 h 52 m, — 60° 11′. (Map 10A).

A beautiful cluster surrounding Kappa Crucis. Splendid in moderate sized telescopes. Visible with binoculars.

NGC 5139, Omega Centauri. 13 h 25 m, — 47° 09′. (Map 8).

A fine globular cluster. Appears as a comet-like object of 4th magnitude to the naked eye.

NGC 5272, M3, Canum Ven. 13 h 41 m, + 28° 32′. (Map 5).

A moderate telescope reveals a beautiful globular cluster.

NGC 5904, M5, Serpentis. 15 h 17 m, + 2° 11′. (Map 6).

A fine globular cluster which appears only as a patch of light with binoculars.

NGC 6205, M13, Herculis. 16 h 41 m, + 36° 31′. (Map 6).

A grand globular cluster just visible to the naked eye. Best seen in large telescopes.

NGC 6341, M92, Herculis. 17 h 16 m, + 43° 11′. (Map 6).

A fine globular cluster resembling M13 though smaller.

NGC 6475, M7, Scorpii. 17 h 52 m, — 34° 48′. (Map 7).

A brilliant open cluster visible with the naked eye.

NGC 6523, M8, Sagittarii. 18 h 02 m, — 24° 23′. (Map 7).

The Lagoon Nebula. Visible with the naked eye. Nebulosity easily seen with binoculars. Rich clusters of stars.

TABLE 7.4
TWENTY DOUBLE STARS

STAR	R.A.	DECL.	MAGS.	SEPARATION	REMARKS (Map reference in brackets)
β TUCANAE	0 h 30 m	—63° 07′	4·5, 4·5	27″	Fine object (10C and D)
ζ PISCIUM	1 h 12 m	+ 7° 25′	4·2, 5·3	24″	Binocular double (1)
γ ANDROMEDAE	2 h 02 m	+42° 12′	3·0, 5·0	10″	Yellow and blue (4)
θ ERIDANI	2 h 57 m	—40° 22′	3·4, 4·4	8″	Fine telescopic double (9)
δ ORIONIS	5 h 30 m	— 0° 18′	2·0, 6·8	53″	Easy with binoculars (3)
μ CANIS MAJORIS	6 h 55 m	—14° 02′	4·7, 8·0	3″	Telescopic object Yellow-blue (3 & 9)
α GEMINORUM	7 h 33 m	+31° 58′	1·6, 2·8	4″	Castor, Telescopic object (3 and 5)
ι CANCRI	8 h 45 m	+28° 52′	4·4, 6·5	31″	Yellow and blue (5)
γ LEONIS	10 h 18 m	+20° 00′	2·4, 3·8	4″	Orange and yellow (5)
α CRUCIS	12 h 25 m	—62° 54′	1·1, 1·9	5″	Acrux, easy telescopically (10A)
α CANUM VEN.	12 h 55 m	+38° 29′	3·2, 5·7	20″	Cor Caroli (5)
ζ URSA MAJORIS	13 h 23 m	+55° 05′	2·1, 4·2	15″	Mizar, naked-eye double with Alcor (Fig. 1.7)
α CENTAURI	14 h 38 m	—60° 44′	0·3, 1·7	4″	Companion is Proxima (10A)
μ BOOTIS	15 h 24 m	+37° 17′	4·0, 6·7	1′ 48″	Easy with binoculars (6)
υ SCORPII	16 h 10 m	—19° 24′	4·2, 6·5	42″	Binocular double (7)
υ DRACONIS	17 h 31 m	+55° 11′	4·6, 4·6	1′ 02″	Easy with binoculars (11)
ε LYRAE	18 h 44 m	+39° 34′	4·6, 4·9	3′ 28″	Each star is double (6)
β CYGNI	19 h 30 m	+27° 53′	3·0, 5·3	35″	Yellow and blue (6)
α CAPRICORNI	20 h 16 m	—12° 37′	3·2, 4·2	6′ 16″	Naked-eye pair (7)
γ DELPHINI	20 h 45 m	+16° 02′	4·0, 5·0	10″	Colour contrast Telescopic (6)

NGC 6656, M22, Sagittarii. 18 h 34 m, — 23° 56′. (Map 7).
A large globular cluster on a par with M13.
NGC 6720, M57, Lyrae. 18 h 52 m, + 33° 00′. (Map 6).
The Ring Nebula. A small telescope will show it as a small oval-shaped blur.
NGC 7092, Cygni, M39. 21 h 32 m, + 48° 18′. (Map 6).
A large open cluster of bright stars. Can be picked up with binoculars.

6. DOUBLE STARS

Table 7.4 contains a selection of double stars, of which the separate components can be seen with the aid of binoculars or a small telescope. A few of them may be split with the naked eye.

TABLE 7.5
SOME WELL KNOWN VARIABLE STARS

STAR	R.A.	DECL.	VARIATION IN MAGNITUDE	PERIOD	REMARKS (Map reference in brackets)
o CETI	2 h 18 m	— 3° 07′	1·7-9·6	331 days	Mira, long-period variable Average max mag about 3·5 (9)
β PERSEI	3 h 03 m	+38° 43′	2·3-3·4	2·9 days	Algol, eclipsing variable, see Whitakers Almanack (4)
α ORIONIS	5 h 54 m	+ 7° 26′	0·1-1·2	—	Betelgeuse, irregular variable (3)
β LYRAE	18 h 49 m	+33° 18′	3·5-4·1	12·9 days	Eclipsing variable (6)
κ PAVONIS	18 h 52 m	—67° 16′	4·0-5·5	9·2 days	Cepheid variable (10D)
η AQUILAE	19 h 51 m	+ 0° 55′	3·7-4·5	7·2 days	Cepheid variable (6)
δ CEPHEI	22 h 27 m	+58° 15′	3·6-4·3	5·4 days	Prototype of the Cepheid variables (11)

7. VARIABLE STARS

Experienced observers study the fluctuations in the brightness of variable stars by comparing them with neighbouring stars whose magnitudes are known. It is exacting work, calling for a great deal of patience. For the beginner, however, it is sufficient to detect the variations in brightness without attempting to draw a light curve, and Table 7.5 gives several variables worth keeping an eye on.

8. OBSERVING THE MILKY WAY

For observers in middle northern latitudes the Milky Way is best seen in midsummer at about midnight when the constellation Cygnus is close to the zenith. On a dark, clear night its ill-defined luminous band can be followed from Cygnus, through the conspicuous W of Cassiopeia, towards the bright yellow star Capella (α Aurigae) low in the north east. The two distinct branches of the Milky Way, separated by the darkness of the Great Rift, can be traced from Cygnus towards the south west, flowing just to the west of Altair (α Aquilae) where it becomes somewhat diminished in brightness. The densest part of the Milky Way in Sagittarius is too low to be seen at its best. Around midnight in December, 10 P.M. in January and 8 P.M. in February, Capella is nearly overhead and the Milky Way is to be found passing through the constellation Aurigae towards the south east just to the east of Orion and the brilliant star Sirius. On the other side of Capella it can be seen in Perseus and Cassiopeia, but Cygnus is now low down in the north west.

Observers in the southern hemisphere, especially those in latitudes between 20° and 40°, are treated to a most glorious view of the Milky Way. At midnight in March, 10 P.M. in April and 8 P.M. in May, the Southern Cross is high in the heavens and Sirius is low in the north west. Near Sirius, the Milky Way is seen as a narrow band, but higher in the sky in the constellation Carina—to the west of the Southern Cross—it becomes much brighter. At this time, Sagittarius and Scorpius are low in the east, but as the night wears on these constellations rise higher in the sky and this section of the Milky Way is exceptionally brilliant. The most magnificent spectacle of all occurs about midnight in June, 10 P.M. in July and 8 P.M. in August, when the Sagittarius cloud is almost overhead. The accumulated light of myriads of stars, interrupted here and there by dark channels of interstellar dust and gas, provides one of the most beautiful naked-eye scenes in the whole of the heavens. Unfortunately, northern observers are denied a satisfactory view of this splendid piece of Nature's handiwork.

No matter where the observer is situated there is great pleasure to be gained from slowly sweeping over the Milky Way with binoculars or a telescope; even in the smallest instrument the field of view is filled with pinpoints of light. Every now and then turn your gaze to a part of the sky at right angles to the Milky Way and note the fall off in the number of stars. For those who do not possess an instrument, star counts can be made quite successfully through a cardboard postal tube.

9. RADIO ASTRONOMY

We have seen how the chance discovery by Jansky of radio noise from outer space marked the beginning of a new and thriving branch of astronomy. By means of large radio telescopes the frontiers of astronomy have been extended far beyond the limits of the most powerful optical instruments, bringing forth fresh discoveries and presenting new problems to astronomers. Why not follow up the brief introduction to radio astronomy given in this chapter by reading other books on the subject? Then again, many large radio observatories are open to the general public, so if you live within a reasonable distance from one why not write to the Director and find out what arrangements there are for visitors.

ANSWERS TO REVIEW QUESTIONS

2. 1980.
6. + 0·49.
7. 0·123 seconds, 26·4 light years.
8. 145 times more luminous.
9. 5012 parsecs.
12. a, b, c and g are galactic.
    d, e and f are extragalactic.

APPENDIX I

ADDITIONAL TIME PROBLEMS

The subject of time was dealt with in Chapter 4. However, the calculations encountered by the ordinary level student warrant further consideration, and the examples given in this section are typical of those set in the examinations.

HOUR ANGLE

The local hour angle (L.H.A.) of a star (or any celestial body) is the arc of the celestial equator, or the angle at the pole, measured from the observer's superior meridian westwards to the hour circle through the star. At upper transit, L.H.A. of a star is 000° or 0 hours, and during the course of a day its value increases to 180° or 12 hours at lower transit, and 360° or 24 hours (0 hours) at the next upper transit. Hour angle measured from the Greenwich celestial meridian is called Greenwich hour angle (G.H.A.). Thus, the difference between L.H.A. and G.H.A. is the longitude expressed in time.

SIDEREAL TIME

At any instant, the local sidereal time (L.S.T.) equals the local hour angle of the First Point of Aries:

$$\text{L.S.T.} = \text{L.H.A. Aries}$$
$$\text{similarly G.S.T.} = \text{G.H.A. Aries}$$

Since difference of hour angle is related to longitude, it follows that:

$$\text{G.S.T.} = \text{L.S.T.} + \text{west longitude in time}$$
$$\text{and G.S.T.} = \text{L.S.T.} - \text{east longitude in time}$$

Sidereal time can be gained from any star by adding its Right Ascension to its local hour angle:

$$\text{L.S.T.} = \text{R.A.} + \text{L.H.A.}$$

A special case occurs when a star is on the observer's meridian, because then the L.H.A. is zero and the L.S.T. is the same as the star's R.A.

Therefore, at upper transit L.S.T. = R.A.

Example 1.

The observed times of two consecutive transits of the star Procyon (R.A. 7 h 37 m 42 s) by a clock which keeps local sidereal time are 7 h 37 m 35 s and 7 h 37 m 33 s. What is the error of the clock and its daily rate?

At upper transit, L.S.T. = R.A.

	1st transit		2nd transit
L.S.T.	= 7 h 37 m 42 s	L.S.T.	= 7 h 37 m 42 s
clock	= 7 h 37 m 35 s	clock	= 7 h 37 m 33 s
error	= 7 s slow	error	= 9 s slow

The clock is, therefore, 9 s slow at the 2nd transit and losing at the rate of 2 s per day.

Example 2.

Find the G.S.T. of the upper transit of the star Arcturus (R.A. 14 h 14 m 14 s) across the meridian of Edinburgh (long. 3° 11′ W).

At upper transit, L.S.T.	=	14 h 14 m 14 s
W. long. in time	=	+ 12 m 44 s
G.S.T.	=	14 h 26 m 58 s

Note: The conversion of longitude into time was made by the method outlined in Chapter 4.

Example 3.

Find the local hour angle of the star Capella (R.A. 5 h 14 m) at Oxford (long. 1° 15′ W) at G.S.T. 6 h 54 m on April 6.

L.S.T. = R.A. + L.H.A. (when a star is not at transit)

Therefore, L.H.A. = L.S.T. — R.A.

G.S.T.	=	6 h 54 m 00 s
W long.	= —	5 m 00 s
L.S.T.	=	6 h 49 m 00 s
— R.A.	=	5 h 14 m 00 s
L.H.A.	=	1 h 35 m 00 s

RELATION BETWEEN MEAN TIME AND SIDEREAL TIME

24 hours of sidereal time = 23 h 56 m 04·1 s of mean solar time.

24 hours of mean solar time = 24 h 03 m 56·5 s of sidereal time.

From these figures it can be shown that for each 1 h of sidereal time the corresponding interval of mean solar time is 9·83 s shorter. Similarly, for each 1 h of mean solar time, the corresponding sidereal interval is 9·87 s longer. For our purposes we can take the difference as 10 s in each case, without incurring any significant error.

The G.S.T. is tabulated in almanacs at 0 hours Universal Time (U.T.) for each day of the year, and the following example shows how this information can be used to obtain the U.T. of upper transit of a star.

Example 4

At 0 hours U.T. on November 16, the G.S.T. was 3 h 37 m 31 s. Find the U.T. on the same date when the star Schedar (R.A. 0 h 36 m 16 s) will make the upper transit of the meridian of Washington (long. 77° 02′ W).

At upper transit, L.S.T.	=	00 h 36 m 16 s
W long.	=	+ 5 h 08 m 08 s
At upper transit, G.S.T.	=	5 h 44 m 24 s
At 0 h U.T. G.S.T.	=	3 h 37 m 31 s
Interval of sidereal time since 0 h U.T.	=	2 h 06 m 53 s

The sidereal interval of 2 h 06 m 53 s must now be converted into a mean solar interval by applying a subtractive correction of 10 s per hour. For this purpose, the sidereal interval may be taken to the nearest 0·1 of an hour, *i.e.* 2·1 h, giving a correction of 21 s.

Interval of sidereal time since 0 h U.T.	−	2 h 06 m 53 s
correction	=	— 21 s
Interval of mean time since 0 h U.T.	=	2 h 06 m 32 s
Thus the U.T. at upper transit	=	2 h 06 m 32 s

Example 5.

The star Procyon (R.A. 7 h 37·7 m) was observed at upper transit of the meridian of Manchester (long. 2° 18′ W) at Universal Time 22 h 20·3 m on February 12. What will be the U.T. of the next upper transit of the star Regulus (R.A. 10 h 07 m) across the same meridian.

Transit of Procyon L.S.T.	= 7 h 37·7 m
Transit of Regulus, L.S.T.	= 10 h 07·0 m
Sidereal interval between transits (diff. of R.A.'s)	= 2 h 29·3 m
correction (2·5 h at 10 sph = 25 s)	= — 0·4 m
Mean solar interval between transits	= 2 h 28·9 m
Transit of Procyon, U.T.	= 22 h 20·3 m
Transit of Regulus, U.T.	= 24 h 49·2 m
i.e.	= 00 h 49·2 m (Feb. 13)

Example 6.

On December 16, the star Betelgeuse (R.A. 5 h 53·5 m) was at upper transit of the meridian of Dover (long. 1° 21′ E) at 0 h 24·7 m Universal Time. At 02 h 08·5 m U.T. a bright star X transited the same meridian.

Find (a) the G.S.T. of the transit of Betelgeuse
(b) the G.S.T. of the transit of star X
(c) the R.A. of X.

(a) Transit of Betelgeuse, L.S.T.	= 5 h 53·5 m
E long. in time (5 m 24 s)	= — 5·4 m
Transit of Betelgeuse, G.S.T.	= 5 h 48·1 m

(b) Transit of Betelgeuse, U.T.		= 0 h 24·7 m
Transit of X,	U.T.	= 2 h 08·5 m
Mean solar interval between transits		= 1 h 43·8 m
correction (1·7 h at 10 sph = 17 s)*		= + 0·3 m
Sidereal interval between transits		= 1 h 44·1 m
Transit of Betelgeuse, G.S.T.		= 5 h 48·1 m
Transit of X,	G.S.T.	= 7 h 32·2 m

* The correction is additive because in this case we are converting a mean solar interval into a sidereal interval, not vice-versa as in examples 4 and 5.

(c) Transit of X,	G.S.T.	= 7 h 32·2 m
E long.		= + 5·4 m
Transit of X,	L.S.T.	= 7 h 37·6 m

Since the L.S.T. of a star's upper transit equals its R.A., it follows that the R.A. of X is 7 h 37·6 m.

EXAMPLES FOR EXERCISE

1. Find the local hour angle of the bright star Sirius (R.A. 6 h 44 m) to an observer at Newcastle upon Tyne (long. 1° 30′ W) when the Greenwich sidereal time is 08 h 30 m.

2. The star Vega (R.A. 18 h 35 m) was observed at upper transit of the meridian of Flagstaff observatory (long. 111° 45′ W) on July 6 at 07 h 08 m universal time. Find the Greenwich sidereal time of transit and also the universal time of transit of Vega at Flagstaff on July 13.

3. At 0 h universal time on September 12, the G.S.T. was 23 h 21 m 15 s. Find the U.T. on the same date when the star Capella (R.A. 5 h 14 m 15 s) will make upper transit of the meridian of Pretoria Observatory (long. 28° 14′ E).

4. The star Vega (R.A. 18 h 35·9 m) was at upper transit of the meridian of Newcastle upon Tyne (long. 1° 30′ W) at 01 h 21·1 m universal time on July 12. The star Deneb (R.A. 20 h 40·4 m) will cross the meridian of Newcastle later. Find the Greenwich sidereal time and the universal time of the upper transit of Deneb.

5. At universal time 0 h on March 6, the Greenwich sidereal time was 10 h 55 m. Find the local hour angle of the star Aldebaran (R.A. 4 h 34 m) at universal time 23 h 00 m on March 6, to an observer at Edinburgh (long. 3° 15′ W).

6. On December 30, the star Rigel (R.A. 5 h 13·1 m) was at upper transit of the meridian of Lowestoft (long. 1° 45′ E) at 22 h 29·4 m universal time. At 23 h 59·9 m universal time a bright star S transited the same meridian.

Find (a) the G.S.T. of the transit of Rigel.
(b) the G.S.T. of the transit of star S.
(c) the R.A. of star S.

ANSWERS

1. L.H.A. 01 h 40 m.
2. G.S.T. 02 h 02 m. July 13, U.T. 06 h 40 m 28 s.
3. U.T. 03 h 59 m 25 s.
4. G.S.T. 20 h 46·4 m U.T. 03 h 25·3 m.
5. L.H.A. 05 h 11 m 50 s.
6. Rigel. G.S.T. 05 h 06·1 m; Star S, G.S.T. 06 h 36·9 m; R.A. 06 h 43·9 m.

APPENDIX II

ASTRONOMICAL QUIZ

1. Name the point on the celestial sphere which is directly overhead.
2. Give the approximate dates when the sun is on the celestial equator.
3. What word describes the slow nodding motion of the earth on its axis?
4. Is the earth closest to the sun in winter or in summer?
5. Is the First Point of Aries in the constellation of Aries?
6. In which constellation is the Pole Star?
7. Which constellation is known as the Northern Cross?
8. Which constellation is known as the Southern Cross?
9. Name the brightest star in the heavens.
10. Give the popular names of Alpha Lyrae, Alpha Bootes and Alpha Canis Majoris.
11. In which constellation is The Plough?
12. Name the three stars which comprise the northern summer triangle.
13. What is the name given to the V-shaped cluster of stars in the constellation Taurus?
14. Name the cluster of stars known as the Seven Sisters.
15. By how much do stars rise earlier each night?
16. What instrument keeps time by the sun we see?
17. Does your watch keep mean time or apparent time?
18. If you journeyed westwards would you advance or retard your watch?
19. What time system do navigators and astronomers use?
20. Who was the first British Astronomer Royal?
21. Who first used a telescope in astronomy?
22. Who invented the reflecting telescope?
23. Which astronomical body is the main cause of tides?
24. What is the interval between successive high tides?
25. Who introduced the leap year?
26. Who gave us the calendar we use today?
27. What is a tropical year?
28. When was the first artificial satellite launched, and what was its name?

29. Where and what are the Van Allen belts?
30. Where in the earth's atmosphere do aurorae occur?
31. What is the average pressure exerted by the atmosphere at the earth's surface?
32. Does the moon possess an atmosphere?
33. Who first set foot upon the moon?
34. At what point in a satellite's orbit is it nearest to the earth?
35. Where is the largest optical telescope in the world?
36. Name the major planets in order of distance from the sun.
37. Between which planets is the main asteroid belt situated?
38. Who first suggested that the moon "falls" around the earth?
39. Name two lunar craters which exhibit a ray system.
40. In which maria did the first manned moon landing take place?
41. What features of the sun, not normally visible, can be seen at a total eclipse?
42. What are Baily's Beads?
43. What is a lunation?
44. When is the full moon on the meridian local time?
45. On average by how much does the moon rise later each day?
46. When does the harvest moon occur?
47. Name three lunar mountain ranges.
48. Seen from the moon, what phase is the earth when the moon is gibbous as seen from the earth?
49. How much of the moon is capable of being seen from the earth?
50. Which spacecraft first photographed the far side of the moon?
51. Who were the first men to see the far side of the moon?
52. When is earth shine most likely to be seen?
53. Does the moon move eastwards or westwards among the stars?
54. Where was the sun situated according to the geocentric system of Ptolemy?
55. Who first suggested that the sun is at the centre of the planetary system?
56. Who stated that the planets move around the sun in elliptical orbits?
57. Who first observed the changing phases of Venus?
58. What is meant by retrograde motion?
59. Can Venus be seen on the meridian at local midnight?
60. When is Venus at its brightest?
61. When is the best time to look for Mercury?

62. Who discovered the "canals" of Mars?
63. When is the most favourable time to observe Mars?
64. Which spacecraft have transmitted photographs of the Martian surface back to Earth?
65. Name the four bright satellites of Jupiter.
66. Where and what is Cassini's division?
67. When did the most recent discovery of a satellite of a planet take place?
68. Name the three planets known respectively as the "winged messenger", the "red planet", and the "noble planet".
69. Do superior planets exhibit phases?
70. Who discovered the planet Uranus?
71. Why are the Leonids so called?
72. Does a comet possess a tail at aphelion?
73. What is the distinction between a meteor and a meteorite?
74. Name the smallest and the largest of the planets?
75. Give two instances in which the correctness of Newton's law of gravitation was verified.
76. Of the known periodic comets name the two having the longest and shortest periods.
77. Why is Halley's comet so called?
78. What is the name given to the visible surface of the sun?
79. What is the solar wind?
80. Of which element is the sun mainly composed?
81. Towards which constellation is the sun's space motion taking place?
82. Where approximately is the sun situated within the Galaxy?
83. What is the standard distance used for the purpose of comparing absolute magnitudes?
84. What are the values of the sun's apparent and absolute magnitudes?
85. Name two red and two white stars?
86. Give the date of the supernova which gave rise to the Crab Nebula.
87. Which celestial objects correspond to the Messier numbers M1, M13, M31 and M42?
88. Name the most distant object visible with the naked eye.
89. Why are planetary nebulae so called?
90. Where is the world's largest fully steerable radio telescope situated?
91. What is the distance of the nearest star to the earth?
92. What is a light year?

93. What is the name and the distance of the nearest star to the sun?
94. Who first demonstrated that the earth rotates on its axis?
95. What evidence is there that the earth revolves around the sun?
96. Towards which constellation is the most dense section of the Milky Way?
97. Why did the early Arabs call Algol the Demon Star?
98. Name in order the colours of the spectrum of visible light.
99. Why are cepheid variables so important to astronomers?
100. Name the nearest extragalactic objects.

Appendix III

ADDITIONAL READING

Encyclopaedia of Astronomy	*Hamlyn*	LAROUSSE
Naked-Eye Astronomy	*Lutterworth*	MOORE
Astronomy with Binoculars	*Faber*	MUIRDEN
Objects for Southern Telescopes	*Camb. U.P.*	HARTUNG
A Field Guide to the Stars	*Collins*	MENZEL
Practical Astronomy	*Laurie*	SCHROEDER
Astronomy	*Van Nostrand*	BAKER
The High Firmament	*Leicester U.P.*	MEADOWS
The Sun and its Influence	*Routledge*	ELLISON
Space Science	*Holt*	TRINKLEIN & HUFFER
Astronomy Old and New	*Bell*	BEET
The Planet Mercury	*Faber*	SANDNER
The Planet Venus	*Faber*	MOORE
A Guide to Mars	*Muller*	MOORE
The Planet Jupiter	*Faber*	PEEK
The Planet Saturn	*Faber*	ALEXANDER
The Sun and the Amateur Astronomer	*Lutterworth*	BAXTER
Our Emerging Universe	*Muller*	BROMS
The Violent Universe	*BBC Pub.*	CALDER
The Astronomers	*Bodley Head*	RONAN
Their Majesties' Astronomers	*Bodley Head*	RONAN
Six Great Astronomers	*Hamilton*	CROWTHER
History of Rocketry and Space Travel	*Nelson*	VON BRAUN
Space	*Lutterworth*	MOORE
Man on the Moon	*Constable*	MANSFIELD
Telescopes and Observatories	*Weid & N. Educ.*	MOORE
Telescopes for Sky-gazing	*Amphoto*	PAUL
How to make a Telescope	*Interscience*	TEXEREAU
Amateur Telescope Making 1, 2 & 3	*Scientific American Pub.*	INGALLS ED.
Outer Space Photography	*Amphoto*	PAUL
The Story of Jodrell Bank	*Hutchinson*	PIPER
History of the Telescope	*Griffin*	KING
History of Astronomy	*Leisure Arts*	REICHEN
Whitaker's Almanack	*Whitaker*	
Norton's Star Atlas	*Gall and Inglis*	NORTON
Scientific American	*Scientific American Pub.*	MONTHLY MAGAZINE
Sky and Telescope	*Sky Publishing Corp.*	MONTHLY MAGAZINE

INDEX